PROBLEM SOLVING

with | Time and Money

Doreen Nation
Sheila Siderman

Creative Publications®

Acknowledgments

Editorial Services Siderman Nation Publishing Services Inc.
Project Editor Janet Pittock
Design Director Karen Stack
Design O'Connor Design
Cover Illustration Lisa Manning
Illustration Jim Dandy
Production O'Connor Design
Manufacturing Dallas Richards

Two Prudential Plaza
Chicago, IL 60601

ISBN 0-7622-1251-9

Catalog #10870

Customer Service 800-624-0822 or 708-385-0110

http://www.creativepublications.com

1 2 3 4 5 6 7 8 ML 06 05 04

Contents

Overview

What Are the Goals of **Problem Solving with Time and Money?**

Throughout their lives, students have a need for working with time and money, in school and out of school, with real-life applications. **Problem Solving with Time and Money** provides students with many situations that represent these types of problems.

Students using **Problem Solving with Time and Money** will learn

- to recognize the relationships among units of time (minutes, hours, days, weeks, months, years).
- to use concepts involving elapsed time and future time to solve problems involving schedules.
- to recognize that understanding elapsed time is essential to everyday living.
- to make change in a variety of situations.
- to compare money amounts and find differences between costs in order to determine the best buy when shopping.
- to use a budget and figure out how long they need to save to achieve a goal.
- that logical reasoning can be applied to solve number puzzles involving money amounts.
- that problems involving time and money can often be solved using problem-solving strategies students already know.

One of the goals of all mathematics education is for students to become successful problem solvers. Toward this end, **Problem Solving with Time and Money** places equal emphasis on content and process. Problems are designed so that in many cases students will be able to choose from a variety of approaches. Often, there is no single best approach. We recommend that as problems are solved, all student suggestions for solution methods be considered. This will reinforce students' understanding that various approaches can often be used to solve the same problem. It will also encourage students to feel comfortable making suggestions. Students are more likely to participate if they understand that incorrect solution paths will be valued for what can be learned from them.

This program is intended for use by second and third grade students who have mastered the basic ideas of addition, subtraction, and have some understanding of multiplication and division concepts. It is assumed that students have studied time and money topics, such as reading calendars, schedules, and clocks; a.m. and p.m. concepts; understanding elapsed time; recognizing coins and bills, understanding their values and making change. Therefore, the program does not teach time and money topics, but rather provides problem situations in which students can apply their understandings to the concepts.

How Is **Problem Solving with Time and Money** Organized?

Sections

Each of six types of problem situations is presented in its own section: *Calendar Math, Elapsed Time, Making Change, Shopping, Budgets,* and *Money Puzzles.*

Each section includes integrated information about instruction, applications, and assessment. Within each section there are three levels of application activities. In order of increasing difficulty, they are Try It Out, Stretch Your Thinking, and Challenge Your Mind.

Instruction

Each section begins with a two-page teacher-oriented Introduction. There is a discussion of the types of problems in the section and the mathematics that students will encounter. This is followed by notes on various problem-solving strategies and solution paths that can be used for the problems in the section. Each introductory section also contains five Thinking About... problems that focus students' attention on number sense and estimation as they apply to the problems they will encounter.

The first problem in each section (Solving Problems with...) will focus students' attention on real-world problems that can be solved using that particular skill. Throughout the program, a variety of strategies and solution paths are presented in these problems.

Applications
The level of sophistication of the exercises increases from one set to the next. Try It Out presents problems in which basic understandings of the section concept are used. Stretch Your Thinking invites students to look beyond basic applications to more sophisticated use of the concept. Challenge Your Mind presents fewer, more complex problems than the previous two applications. Wrap It Up is an application in which students either apply what they have learned throughout the section or use data to create a problem based on the focus of the section.

Assessment
There are multiple opportunities for assessment throughout **Problem Solving with Time and Money**. Each Try It Out, Stretch Your Thinking, and Challenge Your Mind page contains a suggestion for informal assessment. These are in the form of questions to pose to students in order to focus their attention on one important aspect of the problem-solving process.

Each Wrap It Up contains an assessment rubric that sets a performance standard for that activity. You may wish to review the rubric with students before they complete the Wrap It Up. This will let them know what is required to demonstrate a satisfactory level of understanding. Wrap It Up problems can become part of students' math portfolios.

How to Use **Problem Solving with Time and Money** in Your Classroom

Difficulty Level of the Problems
The word problems in this book have been written to appeal to a wide range of student abilities and interests. The Solving Problems with... page and the Wrap It Up activities are designed to be accessible to all students.

Using the Student Pages
Many of the problems can be approached by students working either independently, with a partner, or in a cooperative group. The Solving Problems with... problem is designed to be a teaching problem, presented as a teacher-led activity. You may also find that selected problems or sections are particularly appropriate at a given time, such as when a topic relates well to your daily math lessons.

- **Problem of the Day** If students possess the prerequisite skills needed for a particular section, the Thinking About... problems are particularly suited to this approach. Have students solve one problem daily as a warm-up for any lesson.
- **Problem of the Week** More challenging problems from Stretch Your Thinking or Challenge Your Mind can be presented at the beginning of the week. You might post the problem on a class bulletin board. Students can work the problem during the week, asking you or other students for help as needed any time, up until the problem is due. Solutions can be discussed as a whole-class activity and perhaps posted on the bulletin board with the original problem.
- **Partners or Cooperative Groups** Students can work one or more problems in a section, arrive at solutions that group members agree upon, and present their solutions to the rest of the class. Disagreements over solutions can be worked out either within groups or as a whole class.
- **Whole Class Lesson** In conjunction with studying a topic that is included in **Problem Solving with Time and Money**, you may wish to use a particular page as your math lesson for the day. Problems from the remainder of the page can be assigned as independent work or as a homework assignment.

Using the Teaching Pages
The main teaching technique in **Problem Solving with Time and Money** is the use of thought-provoking questions asked at the appropriate time. Much of the benefit to be gained from the program comes from students recognizing good questions. Over time, this will lead them to begin to ask themselves the kinds of questions needed to analyze a problem. However, this is a lengthy process and you should not look for results in the short run. A commitment to using the program and its approach to problem solving will lead to success in the long run.

Name ______________________________

Problem-Solving Techniques

The Ten Solution Strategies

1. **Act Out or Use Objects** Acting out or using objects can help you show the problem. You can see how the answer fits the problem.

2. **Make a Picture or Diagram** Drawing a picture or simple diagram can help you see all the parts of the problem.

3. **Use or Make a Table** Tables help you keep track of data and see patterns.

4. **Make an Organized List** Making a list makes it easier to look at the data. Seeing the data can help you think step-by-step.

5. **Guess and Check** Guess the answer and check to see if it is correct. If your guess is wrong, use what you learned to make a better guess.

6. **Use or Look for a Pattern** Finding a pattern can help you predict what comes next.

7. **Work Backward** Sometimes you can start with the data at the end of the problem. Using that data to work backward can help you find the answer.

8. **Use Logical Reasoning** Using careful step-by-step thinking can help you solve problems. Sometimes these problems have words such as "if this is true, then...." Making a table often helps to solve logical reasoning problems.

9. **Make it Simpler** When a problem looks very hard, you can use easier numbers and try to solve that problem first. Or you can use fewer items and solve that problem first. Then go back to the original problem and use the same kind of thinking.

10. **Brainstorm** When you cannot think of a way to solve a problem, it is time to brainstorm. Stretch your thinking and be creative.

Name ______________________

Problem-Solving Recording Sheet

Page: _____ Problem Number: _____

Find Out	○ What question must you answer? • What information do you have?
Make a Plan	○ How will you use the information you have to find the answer?
Solve It	• Find the answer. Use your plan. Keep a record of your work. Answer: ______________________
Look Back	○ Does your answer make sense? • How could you check your answer?

1 CALENDAR MATH

Introduction

This section focuses on calendar math. Students work with units of time, ranging from a day to a year, to solve a number of problems. You may wish to have students use the generic calendar found on page 74 as they solve calendar math problems.

Understanding Calendar Math Problems

As students prepare to solve calendar math problems, they will need to consider locating dates and days, scheduling, converting units of time, and calculating elapsed time. Consider the following problems:

How many Fridays are there in February?
To solve this problem, students will need to:

- Use a calendar to locate the month and the correct column for the day indicated.
- Count the number of Fridays.

Cheryl is 105 days older than Ronnie. How many weeks is that?
In a problem such as this, students must:

- Convert days to weeks.

Every evening, except on Tuesdays, Allan spends 1 hour building a dollhouse for the hospital. If the dollhouse takes 41 hours to complete, how many weeks will it take Allan to finish?
In a problem such as this, students must:

- Understand that this is a multistep problem involving finding the number of building hours in each week and then the number of weeks to total 41 building hours.

As students work with calendar math problems, ask these key questions as needed:

- What units of time are in this problem?
- How will using a calendar help you to solve it?
- Does it have multiple steps?
- Is there more than one answer?

Solving Calendar Math Problems

Your students will be more successful if they can apply logical reasoning and combine general problem-solving strategies with methods specific to calendar math such as using a calendar, *use or look for a pattern*, and *guess and check*. Refer to page *vi* for a discussion of strategies.

Solving a Problem by Reading a Calendar
Recognize how a standard calendar is organized.

Example: What day of the week is July 4?

Students must find the month of July on a calendar, identify the fourth day, and use the column heads to identify on what day of the week the date falls.

Solving a Problem by Converting Units Understand how to convert days, weeks, months, and years.

Example: A bear cub grows inside its mother for 180 to 240 days before it is born. How many months is that?

A month is about 30 days; $30 \times 6 = 180$ (or add six 30s); $30 \times 8 = 240$ (or add eight 30s). It is 6 to 8 months.

Solving a Problem by Making an Organized List
Recognize when a list of days and events can be useful.

Example: At Camp Kinooki, pasta is always served on the days before and after the weekend, pot roast on Sunday, pizza on Wednesday, and sandwiches on the other days. On which days are sandwiches served?

Use the strategy *make an organized list* by listing the days in order.

Sunday	pot roast
Monday	pasta
Tuesday	sandwich
Wednesday	pizza
Thursday	sandwich
Friday	pasta
Saturday	sandwich

Sandwiches are served on Tuesday, Thursday, and Saturday.

Assessment

✓ **Informal Assessment** A suggestion for informal assessment will be found on each Try It Out, Stretch Your Thinking, and Challenge Your Mind page. The recommended question will help focus students' attention on one part of the problem-solving process.

Assessment Rubric An assessment rubric is provided for each Wrap It Up. Students' completed work may be added to their math portfolios.

Thinking About Calendar Math

These problems will help students use their number sense and estimation skills as preparation for solving calendar math problems. Present one problem a day as a warm-up. You may choose to read the daily problem aloud, write it on the board, or create a transparency.

1. **Suppose March begins on a Sunday. Can the month also end on a Sunday?**
(No; March has 31 days, 3 days more than 4 full weeks. It ends on the day of the week two days after the day on which it begins.)

2. **A school term lasts 80 weekdays. Does a term take place over a period of 3 months or 4 months?**
(4 months; There are 5 school days in a week and around 20 school days per month. Three months is approximately 60 school days, not including holidays.)

3. **Some people think that a 1-year-old dog would be 7 in human years. If a dog is 40 months old, is it older than 30 human years?**
(No; 40 months can be estimated as about 3 years; 3×7 would be about 21 human years.)

4. **The piano was invented in Italy in 1709. Is it more than 250 years old?**
(Yes; quick mental math reveals that it was 250 in 1959.)

5. **If you sleep 10 hours a night, do you sleep more than 74 hours a week? Explain.**
(No; 10 hours per day $\times$ 7 days per week = 70 hours per week, $70 < 74$.)

1 CALENDAR MATH

Solving Problems with Calendar Math

This lesson will help students focus on the kinds of questions they need to ask in order to solve calendar math problems.

Using the Four-Step Method

Find Out

- ○ The problem asks students to find today's date, the date 2 weeks ago, and the date 10 days from now. Encourage students to describe the problem in their own words so they understand that there are several steps involved in discovering what the days might be.
- ● Maybe; it depends on today's date and whether 2 weeks ago or 10 days hence is in this month or another month.
- ○ Students may estimate that the date 2 weeks ago will fall on the same day of the week. Ten days may be a simple addition problem unless the sum of the dates is greater than the number of days in this month.

Make a Plan

- ● If you count backward twice by a week at a time, you will be pointing to the date 2 weeks ago.
- ○ If today is the 15th or later in the month, students can subtract 14 from today's date. Students can add 10 to today's date but may get a sum greater than the number of days in this month.

Solve It

- ● Answers will vary depending on today's date, the number of days in this month, and the number of days in last month.

Look Back

- ○ Have students compare their results with their estimates. Ask: Is your solution reasonable when compared with your estimate? How did estimating help you solve the problem?
- ● A way to check the solution is to solve by a different method. Encourage students to share their solution methods in small groups.

Name ____________________

1 CALENDAR MATH

Solving Problems with Calendar Math

What is today's date?
What was the date 2 weeks ago?
What will the date be 10 days from now?

Find Out	○ What questions must you answer? • Do you need to know how many days there are in this month? In last month? Why or why not? ○ What is a reasonable estimate?
Make a Plan	• How could counting back help you find an answer? ○ How could addition or subtraction help you find an answer?
Solve It	• Find the answer. Use your plan. Keep a record of your work.
Look Back	○ How does your answer compare with your estimate? • How could you check if your answer is right?

1 CALENDAR MATH

Try It Out

1. Make a Plan To help students plan their solution, ask:

- What does the problem ask you to find? (Find the days on which Mr. Homer wears a green shirt.)
- How can making a calendar help you solve the problem? (Make a week's calendar listing the color of shirt Mr. Homer wears each day.)
- How many days do you need to show on your calendar? (Seven days, so that each day of the week appears once.)
- How can you label the days to show your answer? (Possible answer: Write the name of the appropriate color on the space for each day.)

Solution Path

Mr. Homer wears a green shirt on Saturdays and Wednesdays. Sample calendar:

Sunday	Monday	Tuesday	Wednesday	Thursday	Friday	Saturday
white	blue	red	green	red	blue	green

2. Answers will vary depending on the year; using the generic calendar (page 74), 11 weeks begin in one month and end in another.

3. Answers will vary depending on the year; using the generic calendar (page 74), 3 begin on a Thursday; 2 begin on a Sunday; February, March, and November begin on Thursday; April and July begin on Sunday.

✓ **Informal Assessment** Ask: On which day of the week did this year begin? (Answers will vary depending on the year.) Are there any months that start or end on a Saturday? (Answers will vary depending on the year.)

4. 32 weeks

Students should recognize that school is in session only 5 days each week. Students can count the weeks on a calendar or divide
160 days ÷ 5 days per week = 32 weeks.

5. Answers will vary depending on the year; on the generic calendar (page 74), there are 26 even-numbered and 27 odd-numbered Mondays.

On the generic calendar, Monday, December 31, shares a space with another date. Students may need help understanding this representation.

6. Answers will vary depending on the year; using the generic calendar (page 74): Presidents' Day, Feb. 19; Memorial Day, May 28; Labor Day, Sept. 3; Thanksgiving, Nov. 22.

7. Answers will vary depending on the year; subtract 1965 from this year to find the age of the invention.

Name ____________________

1 CALENDAR MATH

Try It Out

1. Mr. Homer wears a blue shirt on the day before and the day after the weekend. He wears a white shirt on Sunday. He wears a red shirt on days that begin with *T*. He wears a green shirt on the other days. On what days does he wear a green shirt?

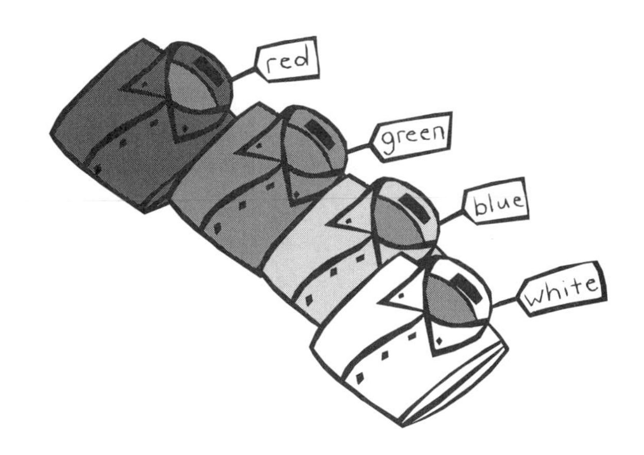

2. How many weeks in this year begin in one month and end in another month?

3. How many months in this year begin on Thursday? On Sunday?

4. Some private schools have classes for 160 days of the year. How many weeks of school is that?

5. How many Mondays this year are on even-numbered days? On odd-numbered days?

6. Some holidays fall on different dates every year. What are the dates of these holidays this year?

   a. Presidents' Day
   3rd Monday in February

   b. Memorial Day
   Last Monday in May

   c. Labor Day
   1st Monday in September

   d. Thanksgiving
   4th Thursday in November

7. James Russell invented the CD in 1965. He got 22 patents for it. How old is his invention?

1 CALENDAR MATH

Stretch Your Thinking

1. Find Out To help students understand the problem, ask:

- What questions does the problem ask you to answer? (How many times a year Robin and Harold will bake for the nursing home if they bake on the second and fifth Saturdays of the month; how many times if they baked on the first and third Saturdays.)
- Would it help to know any of the following:

a. The number of months in the year that had 5 Saturdays? (yes)

b. The number of months in the year with 30 or more days? (no)

c. How many months in the year begin on Sunday, Monday, Tuesday, or Wednesday? (Yes; this helps eliminate months because no month beginning on one of these days can have a fifth Saturday.)

- If you know the number of months with 5 Saturdays, would you still need a calendar to solve the problem? (No, because every month must have at least 4 of each different day of the week.)

Solution Paths

a. 16 times, if you use the generic calendar (page 74); 5 Saturdays in 4 months (March, June, September, December). Every month has a second Saturday. 12 second Saturdays + 4 fifth Saturdays = 16 times baking

b. 24; they would bake twice a month, every month. $2 \times 12 = 24$

Note: If another calendar is used, major holidays, such as New Year's Eve/Day may fall on a baking day, and students may want to give Robin and Harold a day off.

2. Answers will vary; multiply age in years by 12 and add months since last birthday to find age in months.

3. Chicken, April 8 or 9
Duck, April 7 (or April 8) to April 21 (or April 22)
Parakeet, April 3 (or April 4) to April 6 (or April 7)

The answers take into account whether students count the day on which the eggs were laid.

4. Monday to Friday for a 9-day vacation

✓ **Informal Assessment** Ask: How could Vana's mom stretch her week's vacation to 10 days? (Choose a week in which the final weekend is followed by a holiday, such as Labor Day.)

5. about 4 times; round 365 to 360, add $90 + 90 + 90 + 90 = 360$

Students need to recognize that Earth takes 365 days to orbit the sun. A more exact length of Mercury's year is 88 Earth days. Challenge students to use this figure instead of 90 days. Some students may recognize that whether they use 90 or 88 days, the answer will be the same.

Name ____________________

1 CALENDAR MATH

Stretch Your Thinking

1. Robin and Harold bake bread for a nursing home. They bake the second and fifth Saturdays of every month.

   a. How many times this year will they bake for the nursing home?

   b. What if they bake on the first and third Saturday of every month? How many times would they bake this year?

2. What day is your birthday? How many months old are you?

3. Baby birds spend different lengths of time in their eggs before they hatch. Several birds have laid eggs on March 18. Tell what date or dates each might hatch.

Kind of Bird	Days in Egg
Chicken	22
Duck	21 to 35
Parakeet	17 to 20

4. Vana's mom works weekdays. She can take off 5 work days in August. The family wants their vacation to be as long as possible. Which days should she take off?

5. Mercury orbits the sun in about 90 days. About how many times does Mercury orbit the sun in 1 Earth year?

1 CALENDAR MATH

Challenge Your Mind

1. Find Out To help students understand the problem, ask:

- What do parts **a, b,** and **c** of the problem ask you to do? (Find the range of dates on which a whale calf is born; compare the gestation periods of a whale calf and a human infant.)
- What units of time will you use? (**a**, year; **b**, year, month, and day; **c**, month)
- How can finding the answer to **a** help you with **b**? How can finding the answers to **a** and **b** help you with **c**? (Students can subtract 365 days from 547 days, work with 182 days and a starting date of the answer from **a**. Students can count the number of months and fractions of months between January 14 of the first year in the generic calendar and their answers in **a** and **b**, respectively, and subtract 9 months to find the range for the answer in **c**.]

Solution Paths

a. January 13. (leap year, January 12). Determine that 365 days after a given date is the day with the immediately preceding date in the next calendar year.

b. July 14 (leap year, July 13).
547 – 365 = 182 days;
subtract 18 days for January, 182 – 18 = 164;
subtract 28 days for February, 164 – 28 = 136;
subtract 31 days for March, 136 – 31 = 105;
subtract 30 days for April, 105 – 30 = 75;
subtract 31 days for May, 75 – 31 = 44;
subtract 30 days for June, 44 – 30 = 14.

c. 3 to 9 months longer; the earliest date, 1 year, is 12 months; the latest date is 547 ÷ 30 = 18 months; subtract 9 months from each answer. A whale baby takes about 12 to 18 months to develop.

---

2. a. 28-day month, 252 hours; 29-day month, 261 hours; 30-day month, 270 hours; 31-day month, 279 hours; multiply 28 days × 9 hours each day = 252 hours and then add on 9 hours for each additional day.

b. Answers are given in the chart below.

Number of Saturdays in Month	Number of Days in Month			
	28 days	**29 days**	**30 days**	**31 days**
4	256 hr	265 hr	274 hr	283 hr
5	–	266 hr	275 hr	284 hr

Informal Assessment Ask: Can the month of February have 5 Saturdays? Explain. (Yes, but only during a leap year and if Feb. 1 begins on a Saturday; 28 days is 4 full weeks, so in a nonleap year there will be only 4 of each individual day of the week.)

Name ____________________

1 CALENDAR MATH

Challenge Your Mind

1. It takes from 365 to 547 days for a whale calf to be born. A calf began to grow in its mother on January 14.

a. What is the earliest date on which it might be born?

b. What is the latest date on which it might be born?

c. A human baby takes about 9 months to be born. How much longer does a whale take to be born?

2. Suppose you sleep 9 hours every night.

a. How many hours will you sleep this month?

b. How will this number change if you sleep an hour later every Saturday morning? (Remember: Not every month has the same number of Saturdays.)

1 CALENDAR MATH

Wrap It Up

Let's Have a Party!

Discuss the problem with the class before the begin to work on their own. Ask:

- How can you use the information in the problem to find whether there is a day that works? (Eliminate the Saturdays that do not fit the rules and see if there are any days left for this year. Then check those Saturdays in the following year.)

Solutions

Yes, the plan is possible; for 2001 and 2002, the only possibility that fits all the rules is to have the party on February 17th in 2001 and February 16th in 2002.

Solutions will depend on the calendar year. You may wish to give students different calendar years to work with.

Encourage students to share and compare their findings in groups.

Possible Solution Paths

For 2001 and 2002: Students may begin by singling out the first and third Saturdays in February in both years (rule 1). Rule 2 eliminates the first Saturday in the second year, thus calling rule 3 into effect and eliminating the combination of February 3 this year and February 2 next year. The third Saturday in both years, however, fulfills all requirements.

Reading the sample calendars gives the dates February 17 in the first year and February 16 in the second year.

Sample Calendar, 2001

S	M	T	W	T	F	S
				1	2	3
4	5	6	7	8	9	10
11	12	13	14	15	16	17
18	19	20	21	22	23	24
25	26	27	28			

Sample Calendar, 2002

S	M	T	W	T	F	S
					1	2
3	4	5	6	7	8	9
10	11	12	13	14	15	16
17	18	19	20	21	22	23
24	25	26	27	28		

Assessment Rubric

3 The student accurately identifies the first and third Saturdays in February, correctly eliminates the situations that violate the rules, decides whether any date fits all the criteria, and names the correct dates.

2 The student accurately identifies the first and third Saturdays in February, correctly eliminates some of the situations that violate the rules, decides whether any date fits all the criteria, and may name the correct dates.

1 The student accurately identifies the first and third Saturdays in February, does not correctly eliminate the situations that violate the rules, finds only one date that fits the criteria, and names one correctly.

0 The student cannot identify the first and third Saturdays in February, does not correctly eliminate situations that violate the rules, does not correctly decide whether any date fits the criteria, and does not name the correct dates.

Name ____________________

1 CALENDAR MATH

Wrap It Up

Let's Have a Party!

Cal and Mary are planning a party. Here are their rules for choosing the date.

- It must be on the first or third Saturday of February.
- The party cannot be on any of these holidays.

  February 2 — Groundhog Day

  February 12 — Lincoln's Birthday

  February 14 — Valentine's Day

  February 22 — Washington's Birthday
- If the party is on the first Saturday this year, it must be on the first Saturday next year. If the party is on the third Saturday this year, it must be on the third Saturday next year.

Is this plan possible? What dates could they use this year and next year?

2 ELAPSED TIME

Introduction

Elapsed time is the passage of time between a definite starting point and a definite ending point. An understanding of elapsed time is essential for everyday living, which includes reading schedules, keeping appointments, and preparing foods.

Understanding Elapsed Time Problems

To solve a problem involving elapsed time, students will need to determine exactly what they are being asked, what the given information is, what conversions to make, if necessary, and finally to choose one or more methods or operations to find a solution. Consider the following problems:

Jasper is taking part in a walk-a-thon. It is 3:12 p.m., and he and his 13 buddies have been walking for 47 minutes. What time did the walk start?
Before deciding upon a solution method, students will need to:

- Determine what information is essential to solving the problem and what is not. In this problem, for example, the number of buddies with whom Jasper is walking has no bearing on the solution.
- Recognize how using elapsed time can help solve this problem.

Ally has been working on a report about bats for 1 hour 18 minutes. She started at 4:46 p.m. She must stop for dinner at 6:15 p.m. How much longer can she work?
In a problem such as this, students must:

- Understand that this is a two-step problem. First they must determine what time it is now and then they must compare that to the end time. Or they can determine how much time is between 4:46 and 6:15 and then determine how much of that time remains.
- Recognize the relationships between and among the three values given.

As students examine elapsed time problems, ask these key questions as needed:

- What information is given? What do you need to find out?
- Do you need to convert from one unit of measure to another?
- How will using elapsed time help you solve this problem?

Solving Elapsed Time Problems

Your students will be more successful at all levels if they are comfortable with problem-solving strategies such as *work backward*. They may find using a clock or a calendar helpful. Refer to page *vi* for a discussion of strategies.

Solving an Elapsed Time Problem by Adding
Add elapsed time to a given time to reach a solution.

Example: Sylvia started baby-sitting for her little brother at 8:00 a.m. She has been watching him now for 148 minutes. What time is it?

Convert the time that has passed to hours and minutes, and add to the starting time.

148 ÷ 60 = 2 hr 28 min (Students may choose to subtract multiple 60 min increments to convert between minutes and hours instead of dividing.)
8:00 a.m. + 2 hr 28 min = 10:28 a.m.

The time is 10:28 a.m.

Solving an Elapsed Time Problem by Subtracting Subtract elapsed time from a given time to reach a solution.

Example: Celine practices the clarinet for 1 hour and 15 minutes every day. Today she finished at 3:50 p.m. When did she start?

Subtract 1 hr 15 min from 3 hr 50 min.

3 hr 50 min – 1 hr 15 min = 2 hr 35 min

Celine started practicing at 2:35 p.m.

Solving an Elapsed Time Problem by Adding and Subtracting Total several elapsed times and subtract the total from an end time to find a starting time.

Example: Bread dough must rise for 90 min, rest for 10 min, and rise again for 1 hr before baking. If you need to put the bread in the oven at 3:00 p.m., do you want to set it to rise before or after noon?

Add the three amounts of time that pass in the rising process and subtract that total from 3:00 p.m.

90 min + 10 min + 60 min = 160 min = 2 hr 40 min;

3:00 p.m. – 2 hr 40 min = 12:20 p.m.

You need to set the bread to rise at 20 min after noon.

Assessment

✓ **Informal Assessment** A suggestion for informal assessment will be found on each Try It Out, Stretch Your Thinking, and Challenge Your Mind page. The recommended question will help focus students' attention on one part of the problem-solving process.

Assessment Rubric An assessment rubric is provided for each Wrap It Up. Students' completed work may be added to their math portfolios.

Thinking About Elapsed Time

These problems will help students use their number sense and estimation skills as preparation for solving elapsed time problems. Present one problem a day as a warm-up. You may choose to read the daily problem aloud, write it on the board, or create a transparency.

1. **Josh orders a subscription to "Computer Kids" on June 17. He will get the first issue in 3 weeks. Is it possible he will get it before his birthday on July 9?**
   (Yes; there are about 2 weeks left in June; Josh's birthday falls in the second week of July; 4 weeks is more than the 3 weeks it will take to get the first issue.)

2. **Aggie can make 2 kites in 51 minutes. Can she build 6 kites in 2 hours 15 minutes?**
   (No; estimate that 51 min is about 1 hr;
   $6 \div 2 = 3$; 1 hr $\times$ 3 = 3 hr; 3 hr > 2 hr 15 min)

3. **Pedro has promised to read for 2 hours a week. If he reads 15 minutes each day, will he keep his promise?**
   (No; It takes 4 days to reach an hour, so it takes 8 days to reach 2 hours. A week only has 7 days.)

4. **Leroy runs a mile in 5 minutes 13 seconds. Arnold can walk a mile in 22 minutes. If they start together, can Arnold walk 1 mile before Leroy runs 5 miles?**
   (Yes; estimate 5 min per mi $\times$ 5 mi = 25 min; then compare, 22 min < 25 min.)

5. **Jasmine can paint a model car in 30 minutes. It is now 7:20 p.m. Can she paint 5 cars before she goes to bed at 9:30 p.m.?**
   (No; She has 2 hr 10 min until bedtime; estimate that 2 cars take 1 hr, so 5 cars take 2 hr 30 min;
   2 hr 30 min > 2 hr 10 min.)

2 ELAPSED TIME

Solving Problems with Elapsed Time

This lesson will help students focus on the kinds of questions they need to ask themselves in order to solve elapsed time problems. Emphasize that there are several approaches to solving a problem such as *act out or use objects, work backward, make a picture or diagram,* and *make a table.*

Using the Four-Step Method

Find Out	○ To answer the question, students must first answer these questions: How much time did Gabriela have? How much time has she used? How much time is left? Encourage students to describe the problem in their own words to be sure that they understand what they must find out. ● Gabriela has already worked for 1 hr 13 min. ○ Students might estimate by rounding 1 hr 13 min to 1 hr. Subtract the time used (about 1 hr) from the time allowed (2 hr) to estimate the time she has left to work (2 hr – 1 hr = 1 hr).
Make a Plan	● Students may consider using a clock or counting on, coupled with mental math, to find the current time. They may use mental math to find the finish time and then count on or count back to find the difference between the current time and the finish time.
Solve It	○ 47 min Students may use a demonstration clock. They can set the clock hands at the start time and move the minute hand clockwise 1 hr 13 min to find that the current time is 11:13 a.m. They can begin again at the start time and move the minute hand ahead 2 hr to find the finish time of 12:00 p.m. Then they can move the minute hand backward from the finish time to the current time, counting back as they go to find that there are 47 min left for Gabriela to work. Some students may subtract. 2 hr – 1 hr 13 min = 47 min
Look Back	● Have students compare their results with their estimates. Ask them: Does your answer make sense? How does estimating help you solve the problem? ○ Encourage students to check their solutions by using different methods.

Name ____________________

2 ELAPSED TIME

Solving Problems with Elapsed Time

Gabriela began planting peas at 10:00 a.m. She had 2 hours to finish the job. She has been working now for 1 hour 13 minutes. How much time does she have left to work?

Find Out	○ What questions must you answer to solve the problem? • For how long has Gabriela worked so far? ○ How can you estimate the answer to the problem?
Make a Plan	• How will you use the information you have to answer to the question?
Solve It	○ Answer the question. Use your plan. Keep a record of your work.
Look Back	• How does your answer compare with your estimate? ○ How could you check your answer?

2 ELAPSED TIME

Try It Out

1. Find Out To help students plan their solution, ask:

- What does the problem ask you to find? (how long Lin spends working out)
- What part of Lin's workout is not included in the 20 min total? (the time he spends walking)
- Why do you need to know how much time Lin spends walking? (It's part of the time Lin spends on his workout.)
- How could you find out how much time Lin spends walking? (Possible answers: *make an organized list, make a picture or diagram, guess and check*)

Solution Path

Lin's workout will be 30 min long.

Students may make a diagram to solve this problem. Sample diagram:

Jogging — walking
20 min + 10 min = 30 min
jogging — walking — workout

Some students may note that Lin jogs for twice as long as he walks ($4 = 2 \times 2$). They may divide 20 min ÷ 2 = 10 min to find the total amount of walking time and then add 10 min walking + 20 min jogging = 30 min workout.

2. 8 min;
9:16 – 9:08 = 8 min; 9:24 – 9:16 = 8 min;
9:32 – 9:24 = 8 min

Students might count on or count back using a model clock to discover that there is an interval of 8 min.

3. Megan got home 2 min before 6:15 p.m.;
19 min + 31 min = 50 min,
5:23 p.m. + 50 min = 6:13 p.m.

4. 15 min; 3:12 p.m. – 2:57 p.m. = 15 min

5. 12:03 p.m.; 11:46 a.m. + 17 min = 12:03 p.m.

If some students write 12:03 a.m. instead of 12:03 p.m., review how to record the hours before and after noon and midnight.

6. Alex slept 34 min longer than Phyllis.

Students will need to refer to time before and after midnight. Phyllis slept for about 3 hr before midnight and about 7 hr after midnight.
3 hr + 7 hr = 10 hr,
10 hr + 9 min + 5 min =10 hr 14 min
Use the same method to figure Alex's sleep time (10 hr 48 min). Then subtract.
10 hr 48 min – 10 hr 14 min = 34 min

7. about 480 sec; 6:04 – 5:56 = 8 min,
8×60 sec = 480 sec

Students may use repeated addition to solve.

✓ **Informal Assessment** Mars is further from the sun than is Earth. Do you think it takes sunlight more or less time to reach Mars? Explain. (Students should explain that the farther a planet is from the sun, the longer light takes to reach it; so it takes sunlight longer to reach Mars than to reach Earth.)

Name ____________________

2 ELAPSED TIME

Try It Out

1. Lin works out every day. He walks for 2 minutes and jogs for 4 minutes. He repeats this several times. If he jogs for a total of 20 minutes, how much time will he spend on his workout?

2. 

How much time is there between buses?

3. Megan took her puppy for a walk at 5:23 p.m. After 19 minutes, she put him into the basket on her bicycle. She biked for 31 minutes before arriving home. Did she get home before or after 6:15 p.m.? How much earlier or later was she?

4. Paula is a pitcher for her Little League team. She started practicing at 2:57 p.m. Paula threw 48 pitches. She finished at 3:12 p.m. For how long did she practice pitching?

5. Jan is walking to his granddad's house. The trip takes 17 minutes. He leaves his house at 11:46 a.m. What time will he arrive?

6. Phyllis went to bed at 8:51 p.m. She got up at 7:05 a.m. Her brother Alex went to bed at 8:30 p.m. and got up at 7:18 a.m. Who slept longer? How much longer?

7. A ray of sunlight headed toward Earth at about 5:56 a.m. It reached our planet at about 6:04 a.m. About how many seconds does it take for light from the sun to reach Earth?

2 ELAPSED TIME

Stretch Your Thinking

1. Make a Plan To help students decide on a strategy to follow, ask:

- Would it help to know any of the following? Explain.

a. Which member of the family could be ready first? (No; you need to know who will be ready last.)

b. Which member of the family will be ready last? (Yes; this will tell you when the family will be ready to leave.)

c. What is the earliest the family can get to the train station? (Yes; you need to know that they can arrive at 6:55 at the earliest, in order to determine which train they can catch.)

How might making a list help you solve the problem? (A list will help you figure out who came home first, second, and last.)

Solution Path

The 8:15 p.m. train is the earliest train the family can catch.

Sienna's father is home last at 6:15 p.m.;
6:15 + 40 min = 6:55

They will miss the 6:50 p.m. train. The next available train is at 8:15 p.m.

---

2. Jane; Timmy took 7 min + 8 min + 13 min = 28 min,
Jane took 6 min + 9 min + 12 min = 27 min,
27 min < 28 min

Students may see that
7 min + 8 min (Timmy) = 6 min + 9 min (Jane).
So, the problem may be solved by simply comparing 13 min and 12 min.

3. 2:05 p.m.;
11:00 a.m. + 2 hr 25 min = 1:25 p.m.,
1:25 p.m. + 40 min = 2:05 p.m.

4. 7:20 p.m.

Students need to realize that there are 18 children (Max plus 17 friends) and that there is elapsed time *only* before the last 5 sets (the first set of 3 children goes in at 7:00 p.m.).

4 min wait time × 5 groups = 20 min;
7:00 p.m. start time + 20 min. = 7:20 p.m.

Some students might make a list or a table:

7:00 3 children
7:04 6 children
7:08 9 children
7:12 12 children
7:16 15 children
7:20 18 children

5. No, there is only 1 remaining minute;
12 x 10 = 120 sec, 120 sec = 2 min;
3 min × 9 = 27 min, 2 min + 27 min = 29 min;
30 min – 29 min = 1 min

Informal Assessment Ask: If Helen had a 45-minute tape, how many more 3-minute songs could she sing? (She could sing 5 more 3-min songs;
45 min – 29 min = 16 min,
3 min + 3 min + 3 min + 3 min + 3 min = 15 min,
with 1 min remaining.)

Name ____________________

2 ELAPSED TIME

Stretch Your Thinking

1. Sienna's family is going on a train trip. Sienna gets home from school at 3:20 p.m. Her dad gets home at 6:15 p.m. Her mom gets home at 5:40 p.m. It takes 40 minutes to get to the train station. Which is the first train they could take? Why?

Trains Leaving Philadelphia
10:00 a.m.
2:00 p.m.
4:30 p.m.
5:30 p.m.
6:25 p.m.
6:50 p.m.
8:15 p.m.
10:15 p.m.

2. Timmy and Jane left home together. It took Timmy 7 minutes to take muffins to an aunt. Then he took 8 minutes to drop off a video. He got to the park 13 minutes later. Jane took 6 minutes to buy a drink. Then she stopped at her dad's office for 9 minutes. She got to the park 12 minutes later. Who got there first?

3. Jamal flew to Orlando from New York City. His plane left at 11:00 a.m. The flight took 2 hours 25 minutes. It took 40 minutes to get to his aunt's house from the airport. What time did Jamal get to his aunt's house?

4. Max built a maze. He and 17 friends tried it. The first 3 children entered at 7:00 p.m. Another group of 3 entered every 4 minutes. At what time did the last group enter?

5. Helen has a 30-minute tape. She wants to record 12 poems. It takes 10 seconds to read each poem. She also wants to record 9 songs. Each song lasts 3 minutes. Does she have enough time on the tape to sing one more song? Why or why not?

2 ELAPSED TIME

Challenge Your Mind

1. Find Out To help students understand the problem, ask:

- What does the problem ask you do? (Find out how long Julia's lunch break is, how much time Julia spent in science and in art, and how long her school day is.)
- How are hours shown on the military clock? (They are shown in the first two digits of the time. If the number of hours is less than 10, a 0 is put in the first place.)
- How are minutes shown using the military clock? (They are shown in the final two digits of the time.)
- How are the two ways of telling time alike? How are they different? (Alike: the way in which the numbers are written. Different: Students may mention the colon; the a.m. and p.m. designations.)

Solution Paths

a. 50 min; Students may count on or count back to find the answer. They may work with military time or they may convert to civilian time.

b. 85 min or 1 hr 25 min; 50 min + 35 min = 85 min

c. 6 hr 30 min

---

2. Here are two possible rules:

Rule 1 Subtract 12 hours from any time after noon to get civilian time.

Rule 2 If the time is between 1:00 p.m. and 11:59 p.m., add 12 hours. Take out the colon. Take off the p.m.

These rules work because the difference between the two ways of telling time is that civilian time repeats again at twelve noon, which could be interpreted as subtracting 12 hours. The second rule adds the 12 hours back.

Informal Assessment In civilian time, what does the number in a.m. time tell you? (Except for the midnight hour, how many hours and minutes have elapsed since midnight.) What does the number in p.m. time tell you? (Except for the noon hour, how many hours and minutes have elapsed since noon.)

Name ____________________

2 ELAPSED TIME

Challenge Your Mind

The clock that we use is called *civilian time.* Many people in the world use a 24-hour clock. It also may be called *military time.* The chart shows both times.

0100 is 1:00 a.m.	1300 is 1:00 p.m.
0115 is 1:15 a.m.	1340 is 1:40 p.m.
0200 is 2:00 a.m.	1400 is 2:00 p.m.
1000 is 10:00 a.m.	2000 is 8:00 p.m.
1130 is 11:30 a.m.	2300 is 11:00 p.m.
1200 is 12:00 noon	2400 is 12:00 midnight

1. Julia writes her daily schedule in military time.

a. How much time does Julia have for lunch?

b. What is the total time she spends in science and art?

c. How long is Julia's school day?

Julia's Schedule	
0805	Leave for school
0830	Homeroom starts
0845	Music
0935	Science
1025	Recess
1050	Reading
1150	Lunch
1240	Math
1340	Social Studies
1425	Art
1500	Go home

---

2. Work with a partner. Write a rule for using military time. Tell why your rule works.

2 ELAPSED TIME

Wrap It Up

Busy, Busy, Busy

Discuss the task with the class before students begin to work on their own. Ask questions such as:

- Before making your schedule, what might you do? (Make a list of a day's activities.)
- Before you compare the amount of time you and your partner spend on activities, what do you have to do? (Be sure that each activity has a start time listed on the schedule. Use the start times to figure out the time spent on each activity.)
- When might you need to make a conversion before comparing? (One partner may have time indicated in hours; another may use minutes.)
- You may wish to use these questions to help students as they create their own problems:
- (Part **b**) What things did both you and your partner do? What did you do that your partner did not do?
- (Part **c**) How many minutes more did your partner spend on an activity than you did?

Solutions

a. Students might use a demonstration clock to count back or count on to figure the times spent. Students need to identify which activities they can compare and which ones they cannot.

b. Check students' problems. Students need to determine how much time they spent on two (or more) activities and then subtract to determine how much less time they spent on one than the other.

c.–d. Check students' problems. Students' problems should involve some comparison of time spent on the same activity.

Assessment Rubric

3 The student writes an accurate schedule; correctly determines the amounts of time spent on two or more different activities; makes accurate comparisons; and successfully solves a partner's problem.

2 The student writes an accurate schedule; correctly determines the amounts of time spent on two different activities; makes accurate comparisons but has some difficulty comparing amounts; has some success solving a partner's problem.

1 The student writes a schedule but either has difficulty reading the schedule or has limited success in finding the activities that fit the categories; makes comparisons but has difficulty comparing amounts; has partial success solving a partner's problem.

0 The student cannot write a schedule without assistance; has difficulty using the schedule, or cannot make comparisons between two different activities; does some of the work but does not show any understanding of elapsed time.

Name ______________________

2 ELAPSED TIME

Wrap It Up

Busy, Busy, Busy

Work with a partner. Each of you writes down everything you do for one day. Begin with the time you get up in the morning and end with dinner.

a. Compare your schedule with your partner's.

b. Make up a problem about how much less time you spent on one activity than on another.

c. Make up a time problem about an activity that both you and your partner did.

d. Switch problems with your partner. Solve. Then check each other's work.

3 MAKING CHANGE

Introduction

This section focuses on making change. Students are asked to identify various coins and make change with different coins and bills in a variety of circumstances. This section is also designed to build reasoning skills in preparation for the following sections on shopping and budgeting.

Understanding Making Change Problems

As students prepare to solve money problems, they will need to determine the total amount of purchases, how many coins or bills are necessary for a purchase, and how to make change from a given amount. Consider the following problems.

Julie buys a whiffle ball for $1.19 and a plastic bat for $4.97. She gives the clerk a $10 bill, and asks for change mostly in quarters so she can make a phone call. What change would you give her?
To solve this problem, students will need to:

- Find the total cost of the two items Julie is buying.
- Find the difference between the total cost and $10.
- Figure out how many quarters can be used to help make up the difference to give Julie as change.
- Understand that this is a multistep problem, and each step must be done in order.

As students work at making change, ask these key questions as needed:

- What money amounts are given in the problem?
- How are the different amounts of money related to each other?
- Does the problem have multiple steps?
- Is there more than one answer?

Solving Making Change Problems

Your students will be more successful at all levels if they are comfortable with problem-solving strategies such as *work backward*. They may find using play money helpful. Refer to page *vi* for a discussion of strategies.

Solving a Problem by Counting On Counting on from the sale price to the amount of money tendered is one way to arrive at the change.

Example: Angie buys a pizza for $2.97. She gives the clerk a $5 bill. What change should she receive?

Starting with $2.97, students can add on $0.03 to reach $3.00 and then add on $1.00 at a time to reach $5.00.

Her change is $2.03.

Solving a Problem by Counting Back Counting back is another approach to finding the difference between the money tendered and the purchase price.

Example: Julio has one $5 bill, a dollar coin, 3 quarters, 6 dimes, 8 nickels, and 3 pennies. He is selling lemonade for $1.25. Rhoda has only a $10 bill. Can Julio make change for her?

Subtract from $10:
$10 – $5 ($5 bill) = $5
$5 – $1 (dollar coin) = $4
$4 – $1 (nickels and dimes) = $3
$3.00 – $0.75 (quarters) = $2.25
Mentally figure that the pennies are not enough to make the difference.

No, Julio cannot make change for Rhoda.

Solving a Problem by Adding and Subtracting
Adding and subtracting is another way to find the change from a purchase.

Example: Kay wants to buy soy sauce for $2.29, rice for $3.45, and chicken for $4.25. If she pays with a $10 bill, how much change will she get?

$2.29 + $3.45 + $4.25 = $9.99
$10.00 – $9.99 = $0.01

She will get one cent change.

Assessment

Informal Assessment A suggestion for informal assessment will be found on each Try It Out, Stretch Your Thinking, and Challenge Your Mind page. The recommended question will help focus students' attention on one part of the problem-solving process.

Assessment Rubric An assessment rubric is provided for each Wrap It Up. Students' completed work may be added to their math portfolios.

Thinking About Making Change

These problems will help students use their number sense and estimation skills as preparation for solving problems with making change. Present one problem a day as a warm-up. You may choose to read the daily problem aloud, write it on the board, or create a transparency.

1. **Walt wants to buy a can of spinach for 89¢, a loaf of bread for $2.09, and a half gallon of milk for $1.75. If he pays with a $5 bill, will he get any bills in his change?**
   (No; Mentally round up or down to estimate that $1 + $2 is $3; $3.00 + $1.75 = $4.75; $5.00 – $4.75 = $0.25; $0.25 < $1.00.)

2. **Nancy has 18 coins in the cash register, and none are dollar coins. Is it possible that she could change a $10 bill? Explain your thinking.**
   (No; if all the coins were the greatest possible value—a half-dollar—they would only be worth $9.)

3. **Suppose you want to make 67¢ using only two kinds of coins. How could you do that using the fewest coins?**
   (Use 6 dimes and 7 pennies.)

4. **Eliot is charged $2.99 for a snow scraper for the car. He pays with a $5 bill. After he gets change, he realizes that the scraper was supposed to cost $2.37. Should the clerk give Eliot $0.37 or $0.62? Explain.**
   ($0.62; the clerk owes Eliot the difference between the correct price and the price charged.)

5. **Ivy takes 3 coins out of her pocket. Could she have more than $1.50? Tell how you know.**
   (Yes, if at least one of the coins is a dollar coin, and if at least one other coin is a half-dollar.)

3 MAKING CHANGE

Solving Making Change Problems

This lesson will help students focus on the kinds of questions they need to ask themselves in order to solve rate problems. Emphasize that there may be several approaches to solving a problem.

Using the Four-Step Method

Find Out	○ The problem asks students to find the 5 coins the clerk used in making change. Encourage students to describe the problem in their own words so they understand that there are several steps involved in discovering what the coins might be. ● Yes, because that is the total amount of change that the clerk should give. ○ Students may estimate that they will need 5 coins that are about 25¢. Five nickels is close to the right answer.
Make a Plan	● It would simplify the problem. \$3.74 + \$1.00 = \$4.74; \$5.00 – \$4.75 = \$0.26; \$0.26 is needed.
Solve It	○ 1 dime, 3 nickels, and 1 penny Students can use play money to make change. Start with \$3.74. Add on \$1.00. Continue adding on to reach \$5.00, and note that \$0.26 still needs to be added in coins. Start with 1 penny and find 4 other coins that go with it to make \$0.26.
Look Back	● Have students compare their results with their estimates. Ask: Is your solution reasonable when compared with your estimate? How did estimating help you solve the problem? ○ A way to check the solution is to solve by a different method. Encourage students to share their solution methods in small groups.

Name ______________________

3 MAKING CHANGE

Solving Making Change Problems

Kim buys a hat for $3.74 and gives the clerk a $5 bill. The clerk hands him a $1 bill and 5 coins. What coins did Kim get?

Find Out	○ What question must you answer to solve the problem? • To solve the problem, do you need to find the difference between $3.74 and $5.00? Explain. ○ What is a reasonable estimate?
Make a Plan	• How would adding $1.00 to $3.74 help you find the answer?
Solve It	○ Answer the question. Use your plan. Keep a record of your work.
Look Back	• How does your answer compare with your estimate? ○ How could you check your answer?

3 MAKING CHANGE

Try It Out

1. **Find Out** To help students understand the problem, ask:

- What does the problem ask you to find? (What coin did Sue drop?)
- Would it help you to know how much money Sue should give as change? (Yes; one way to solve this is to find the difference between the correct change and the actual change.)

Solution Path

a dime

Students need to recognize that this is a two-step problem. First they need to find how much money Sue should give Paulo. ($5.00 – $2.58 = $2.42) Then they need to find the difference between what she should have given him and what she actually did. ($2.42 – $2.32 = $0.10)

---

2. 3¢; $1.00 – $0.97 = $0.03

Students may mentally calculate this answer.

3. $64; 4 × $1 = $4, 4 × $5 = $20, 4 × $10 = $40, $4 + $20 + $40 = $64

4. 3; $3.00 – $2.25 = $0.75, 3 × $0.25 = $0.75

5. 2 quarters and 1 dime, or 1 half-dollar and 2 more nickels; 25¢ + 25¢ + 10¢ + 5¢ = 65¢ or 50¢ + 5¢ + 5¢ + 5¢ = 65¢

6. 35¢; 40¢ + 25¢ = 65¢, $1.00 – $0.65 = $0.35

✓ **Informal Assessment** Ask: If Nina decided to also buy a 35¢ stamp, how much change would she get? (She wouldn't get any change.)

7. Possible answer: 1 half-dollar and 2 dimes; $2.00 – $1.04 = $0.96, $0.96 – $0.26 = $0.70, $0.50 + $0.10 + $0.10 = $0.70

The total change is $0.96. Many other coin combinations are possible.

8. Yes, he will have 25¢ left; 7 quarters are $1.75, 4 dollar coins are $4.00, $1.75 + $4.00 = $5.75, $5.75 – $5.50 = $0.25

9. the most, $4 (4 dollar coins); the least, 4¢

Name ____________________

3 MAKING CHANGE

Try It Out

1. A school T-shirt costs $2.58. Paulo gives Sue a $5 bill. As she gives him change, she drops one coin. She hands Paulo $2.32. What coin did Sue drop?

2. Randi buys a 97¢ notebook with a $1 bill. How much change will she get?

3. Michael is selling school pens. He has four $1 bills, four $5 bills, and four $10 bills to use for change. How much money does he have?

4. Latoya buys a game. It costs $2.25. She pays with three $1 bills. What is the greatest number of quarters she could get as change?

5. Donald gives 65¢ to the Red Cross. He gives 4 coins. One coin is a nickel. What are the other coins?

6. Nina buys an eraser for $0.25 and a banana for $0.40. She gives the clerk $1.00. How much change should she receive?

7. Hector is buying juice for $1.04. He gives the clerk $2.00. The clerk takes 1 penny and 1 quarter from the cash register. What other coins will she use to make the change?

8. André has 7 quarters and 4 dollar coins. Can he buy a kite for $5.50? If so, how much change will he have left?

9. Ellen takes 4 coins out of her pocket. What is the greatest amount of money she could have in her hand? What is the least?

3 MAKING CHANGE

Stretch Your Thinking

1. Find Out To help students understand the problem, ask:

- What question does the problem ask you to answer? (What three kinds of coins can you use to make 47¢).
- Are there any kinds of coin that would not be useful in making 47¢? (dollar and half-dollar coins)
- Is there any kind of coin you know you must have? (at least two pennies)
- Is this a problem that has only one answer or more than one? (more than one possible answer)

Solution Paths

Students may *work backward, use objects,* or *make a table*:

32 pennies = 32¢; 47¢ – 32¢ = 15¢;
1 dime = 10¢; 15¢ – 10¢ = 5¢;
1 nickel = 5¢; 5¢ – 5¢ = 0

There are more than 20 combinations. Some are listed below.

Q	D	N	P
	1	1	32
	1	2	27
	2	2	17
1	1	1	7
1	2		2

2. one $1 bill and 4 pennies; $0.15;
$1.04 – $0.89 = $0.15

3. 8 coins; 1 half-dollar, 1 quarter, 2 dimes, 4 pennies;
50¢ + 25¢ + 20¢ + 4¢ = 99¢

4. The most, $204—two $100 bills and 4 dollar coins; the least, $2.04—two $1 bills and 4 pennies.

5. 6 bills and 3 coins; one $10 bill, one $5 bill, four $1 bills, 1 half-dollar, 1 dime, 1 penny;

$20.00 – $0.39 = $19.61

$10 + $5 + $4 + $0.50 + $0.10 + $0.01 = $19.61

Informal Assessment Ask: How could you make 61¢ without using half-dollars or dimes? (2 quarters, 2 nickels, 1 penny)

6. $1.00 for the eggs and $1.05 for the milk;
2 × $0.50 = $1.00 for the eggs,
$1.00 + $0.50 for roll = $1.50,
$2.55 – $1.50 = $1.05 for milk

7. $1.04; $16.53 – $15.49 = $1.04

Name ______________________

3 MAKING CHANGE

Stretch Your Thinking

1. How can you make 47¢ using only three kinds of coins? Find three different ways.

2. Aaron doesn't like carrying pennies. He has a $1 bill, 3 quarters, and 4 pennies. He buys a model airplane for 89¢. How can he pay for the plane and get rid of his pennies? What will his change be?

3. Janet has 99¢. What is the least number of coins she could have? What are they?

4. Jack has 2 bills and 4 coins. What is the greatest amount of money he could have under $250? What is the least?

5. Mr. Khan buys a pencil for 39¢. He pays with a $20 bill. What is the fewest bills and coins that he can get as change? What are the bills and coins?

6. You spend $2.55 on groceries. A roll costs $0.50. You also buy a dozen eggs and a quart of milk. The eggs cost twice as much as the roll. How much do you spend on eggs? On milk?

7. Sonia paid $15.49 for a pair of jeans. Then she finds another pair she likes more for $16.53. She wants to make a trade. How much more will she pay?

3 MAKING CHANGE

Challenge Your Mind

1. **Make a Plan** To help students decide on a strategy to follow, ask:

- Name the kinds of coins that might be among those in Evan's bank. (half-dollars, quarters, dimes, nickels, pennies)
- Does the problem place any limits on which kinds of coins he has or how many of any type? (no)

Solution Paths

There are many answers ranging from 13¢ for 13 pennies to $13 for 13 dollar coins. Students can *make a table* or use play money.

---

2. **a.** 12 nickels; Some students may use division, $60 \div 5 = 12$.

b. 6 dimes

c. 1 quarter and 7 nickels or 2 quarters and 2 nickels

d. 2 quarters and 1 dime

e. 1 quarter and 3 dimes and 1 nickel; 1 quarter and 2 dimes and 3 nickels; 1 quarter and 1 dime and 5 nickels

✓ **Informal Assessment** Ask: How could this problem help you with making change? (Possible answer: It makes you think about using different combinations of coins, especially if a limited number of some coins are available.)

Name ___________________

3 MAKING CHANGE

Challenge Your Mind

1. Evan has 13 coins in his bank. Find 5 different amounts of money he could have. Tell what coins make each amount.

---

2. How can you make 60¢ using these coins?

a. Only nickels

b. Only dimes

c. Nickels and quarters

d. Dimes and quarters

e. At least one dime, one quarter, and one nickel

3 MAKING CHANGE

Wrap It Up

Picture Perfect

Before students begin to work on their own, ask:

- What limit is there on how much the shopper can spend? ($10)
- What is the most expensive item for sale at Bart's Art Store? (package of 5 markers for $7.35)
- If the shopper buys one product, how can you figure out the change? (Subtract the cost of the item from $10.)
- If the shopper buys many items, how can you figure out the change? (Add the costs and subtract the total from $10; or subtract from $10, one item at a time.)

Solutions

Answers will vary. Students' problems should fulfill the conditions set forth in the text. Sample responses are given below.

Part 1

- Molly goes into Bart's Art Store and buys some gold paint with a $10 bill. How much change does she receive? ($10.00 – $1.75 = $8.25)
- Mary Elizabeth buys glitter for $3.75. She only has quarters. How many quarters must she give the clerk? (15 quarters)

Part 2

- Levar buys a package of markers and a small pad of paper to make a picture. How much change will he get for $10? (none)
- Shante buys blue paint and a small pad of paper. She gives the clerk $10. How much change does she get? ($6.16; $1.19 + $2.65 = $3.84, $10.00 – $3.84 = $6.16)

Assessment Rubric

3 The student uses the given information to write problems that meet all the criteria and solves the problems correctly.

2 The student uses the given information to write problems that meet the criteria and finds correct solutions for only some of them.

1 The student either does not write problems that meet all the criteria and/or finds the correct solutions for only some of them.

0 The student is unable to use the given information to write problems.

Name ____________________

3 MAKING CHANGE

Wrap It Up

Picture Perfect

Part 1

Bart's Art Store is having a sale. Look at the ad. A shopper with a $10 bill wants to buy one item. Make up a problem about the shopper. Solve it and share it with a friend.

Part 2

A shopper buys several things to make a picture. Make up a problem about this shopper. Swap problems with a friend and solve it. Share your answers.

Bart's Art Store Giant Sale	
Paint Remover	$2.50
Small Brush	$1.86
Big Brush	$5.45
Varnish	$1.24
Gold Paint	$1.75
White Paint	$1.19
Black Paint	$1.19
Green Paint	$1.19
Bright Red Paint	$1.35
Blue Paint	$1.19
Yellow Paint	$1.19
Orange Paint	$1.19
Small Pad of Paper	$2.65
Big Pad of Paper	$6.25
Canvas	$4.19
Package of 5 Markers	$7.35
10 Colored Pencils	$7.25
Glitter in 5 Colors	$3.75
Special: Any 5 Paints	$5.27

4 SHOPPING

Introduction

This section focuses on using money in shopping situations. Students determine costs, which bills and coins make different money amounts, and what change to expect. They also compare prices to figure the better buy.

Understanding Shopping Problems

As students prepare to solve shopping problems, they will need to find the total cost of multiple items, find differences between costs, and make comparisons. Consider the following problems:

Milk costs $2.99 a gal; 1 doz eggs costs $1.89. Archie buys 2 gal of milk and 3 doz eggs. He can use a $2-off coupon with a purchase of 2 gal of milk. How much change will he get from a $10 bill?
To solve this problem, students will need to:

- Find the total cost of the 5 items Archie buys.
- Deduct the amount of the coupon.
- Find the difference between the final cost and $10.00.

A ride on the city bus costs $1.25 and requires exact change. Dena has a $5 bill. What can she buy for breakfast so that she is sure to get the change she needs for the bus?

Bagel and cream cheese	75¢
Doughnut	60¢
Roll and butter	50¢

To solve this problem, students will need to:

Find the difference between the cost of each item and $5.00.

Identify the item that Dena could buy to ensure that she gets the change she needs.

Sue and Sol want to spend as much of their money as possible on gifts for their mom. Together, they have $25. They have found some things their mom would like. What should they buy?

Necklace	$ 7.50
Music Book	$15.00
Car Wax	$ 6.50
T-shirt	$16.00
Pastels	$ 3.50

To solve this problem, students will need to:

- Find the difference between the cost of various item combinations and $25.00.

As students work with shopping problems, ask these key questions as needed:

- What amounts of money are stated in the problem?
- How do the different amounts of money relate to each other?
- Do you have to compare prices?
- Is there more than one answer to the problem?

Solving Shopping Problems

Your students will be more successful if they can apply logical reasoning and combine general problem-solving strategies with methods specific to shopping such as using *guess and check*, using play money, and *use or make a table*. Refer to page *vi* for a discussion of strategies.

Solving a Shopping Problem by Counting On
Counting on from the cost to the amount of money tendered is one way to find the amount of change.

Example: Marcus buys an adventure book for $5.65. He gives the clerk a $10 bill. The clerk gives him four $1 bills, 2 dimes, and 1 nickel. What's wrong?

The clerk gave Marcus 1 dime too few. Starting with $5.65, students can add on $4.00 (for the $1 bills) to reach $9.65, and then add on a nickel ($9.70) and 2 dimes ($9.90). They can mentally calculate the difference between $9.90 and $10.00.

Solving a Shopping Problem by Counting Back
Counting back from the money given to a clerk for a purchase is another approach to finding the difference between the money given and the purchase price.

Example: Mia buys a hammer for $4.30. She gives the clerk a $5 bill. How much change should she get?

Students can count back by dimes, $5.00, $4.90, $4.80, $4.70, $4.60, $4.50, $4.40, $4.30. That's 7 dimes, or $0.70 change.

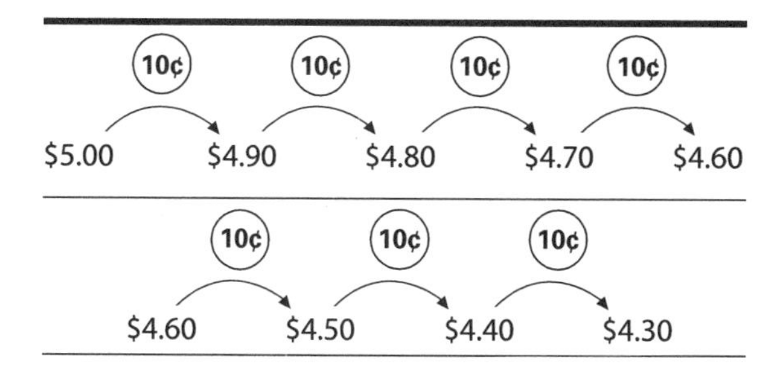

Assessment

Informal Assessment A suggestion for informal assessment will be found on each Try It Out, Stretch Your Thinking, and Challenge Your Mind page. The recommended question will help focus students' attention on one part of the problem-solving process.

Assessment Rubric An assessment rubric is provided for each Wrap It Up. Students' completed work may be added to their math portfolios.

Thinking About Shopping

These problems will help students use their number sense and estimation skills as preparation for solving shopping problems. Present one problem a day as a warm-up. You may choose to read the daily problem aloud, write it on the board, or create a transparency.

1. **A good pillow costs $2.99. A better pillow costs $4.59. The best pillow costs $7.89. Is it cheaper to buy the best pillow or both a good pillow and a better pillow? Explain.**
(Both the good and better pillows; mentally estimate that $3.00 + $4.60 is $7.60; $7.60 < $7.89.)

2. **Hildie wants a soccer ball that costs $12.83. If she has three $1 bills and 10 coins, is it possible that she has enough money?**
(Yes; if all the coins were the greatest possible value, $1, she would have $13.)

3. **Adam wants to buy three items. They cost $0.68, $0.83, and $0.96. He has $1.55. If he puts back the least expensive item, will he have enough money?**
(No; estimate $0.80 + $1.00 = $1.80. Mentally figure $1.80 > $1.55.)

4. **Jenna has $3.00. She wants to buy two different flowers. Can she? Explain.**

Carnation	$1.29
Lily	$1.75
Rose	$2.00
Orchid	$5.00

(No; the least expensive pair can be mentally estimated as costing more than $3.00 since $1.25 + $1.75 would be $3.00.)

5. **A package with 5 bars of soap costs $2.25. A single bar costs $0.50. Which is a better buy?**
(The package with 5 bars; 5 bars of soap bought separately would cost $2.50; $2.50 > $2.25.)

4 SHOPPING

Solving Shopping Problems

This lesson will help students focus on making change. Emphasize that there are several approaches to solving shopping problems. Students might count on to figure costs and to check change. They might use play money or make a table. Sometimes they can simplify by rounding and estimating.

Using the Four-Step Method

Find Out

- ○ The problem asks students to find the amount of change that Allie should receive. Encourage students to describe the problem in their own words in order to be sure that they understand it.
- ● One possible estimation strategy might be to round $0.91 to $0.90 and $1.01 to $1.00 and mentally figure that the change should be a dime.

Make a Plan

- ○ If the problem is change from $1, students should know that $1.00 – $0.91 = $0.09. They may count back, subtract, or use play money.
- ● If students rounded $0.91 to $0.90 and $1.01 to $1.00 to estimate a dime in change, they may be able to use mental math to solve the problem. Other students may count back, subtract, or use play money.

Solve It

- ○ She should get back 10¢. Two possible solution paths are:

Logical Reasoning In the estimation, each figure was rounded down one cent. Therefore, the difference between the rounded numbers is the same as the difference between the original numbers.

Counting On Students may start from $0.91 and count up to $1.01. Or, they may begin with the mental calculation that if Allie had paid $1.00, she would have received 9 cents in change and count on 1 more penny to make 10 cents change.

Look Back

- ● Have students compare their results with their estimates. Ask them to tell whether they think their solutions are reasonable in light of their estimates. Ask whether they think estimating helped them solve the problem.
- ○ Students can check their work by using one of the methods not used to solve the problem. For example, students who based their answers on their estimations may count on using play money to check. Allow time for students to share their solution methods in small groups.
- ● Some students may realize that Allie may have preferred to have a dime to having a nickel and 4 pennies or 9 pennies. Ask students to discuss the different reasons that people might try to alter the amount of change they receive.

Name ____________________

4 SHOPPING

Solving Shopping Problems

Allie is buying a goldfish that costs 91¢. She gives the clerk a $1 bill and a penny. What change should she get back?

Find Out	○ What question must you answer to solve the problem? • How can you estimate the answer to the problem?
Make a Plan	○ How would you solve the problem if Allie had given the clerk $1 for the goldfish? • How will you use the information you have to find the answer to the problem?
Solve It	○ Find the answer. Use your plan. Keep a record of your work.
Look Back	• How does your answer compare with your estimate? ○ How would you check to see whether your answer is right? • Can you think of another way to solve the problem?

4 SHOPPING

Try It Out

1. Find Out To help students understand the problem, ask:

- What question does the problem ask you to answer? (How much will Harris pay for the turkeys?)
- How can you estimate how much Harris will pay? (Students may round the prices and add them. If their estimate is over $25, they may subtract $2.)

Make a Plan To help students decide on a strategy to follow, ask:

- What must you figure out before finding the final solution? (You have to calculate the cost of the two turkeys to see if the amount is over $25.)

Solution Path

$23.17; $10.80 + $14.37 = $25.17,
$25.17 – $2.00 = $23.17

2. $1.73; students may write an addition problem or use coins to count with.

Students may also make the problem simpler by rounding 98¢ to $1, and then subtracting 2¢ from the 75¢. If no students use the approach of completing the dollar, you might want to demonstrate this technique and discuss when it is a useful approach.

3. 1 quarter and 2 dimes, or 1 quarter, 1 dime, and 2 nickels

Students may use play money or calculate mentally.

4. 3 bills (one $5 bill and two $1 bills) and 3 coins (half-dollar, quarter, and nickel); if students consider the $2 bill, then the answer is 2 bills, (one $5 bill and one $2 bill).

Students may use play money to make the combination.

Informal Assessment Ask: Why is it better to make up the dollar amount in bills? (It takes only 3 or 2 bills, but many coins.)

5. the $1.25 item

Students may just choose the most expensive item in the basket. Encourage them to find the total cost of the remaining items to check their answer.

6. two slices of cheese and one of pepperoni

Students need to understand the restrictions: One slice must be pepperoni, which costs $1.35. That leaves John with $1.85. So, he can buy only two slices of cheese pizza.

Name ______________________

4 SHOPPING

Try It Out

1. Harris is buying turkeys for dinner. One costs $10.80. The other costs $14.37. He has a coupon for $2.00 off purchases of $25.00 or more. How much will the turkeys cost him?

2. Patrick gets a cup of yogurt for 98¢ and a can of juice for 75¢. How much does he spend for his snack?

3. Tanya wants to pay the exact amount of her purchase. It costs 45¢. She has 1 quarter, 3 nickels, and 2 dimes. What are two ways she can pay?

4. Angela wants a lizard that costs $7.80. What is the smallest number of bills and coins she needs to have the exact amount? What are they?

5. Patty has four items in her shopping basket. They cost $0.75, $1.25, $0.85, and $0.90. Patty has $2.70. What is the least expensive item she might put back so she can pay with the money she has?

6. Jeff has $3.20 to buy pizza slices for himself, his sister, and his brother. His brother wants pepperoni, his sister will not eat meat, and Jeff doesn't care what he has. What should he order?

4 SHOPPING

Stretch Your Thinking

1. Find Out To help students understand the problem, ask:

- What does the problem ask you to find? (if it is better to buy both a small and medium lemonade instead of a large one)
- What does *better buy* mean? (It means cheaper or getting more for your money or what the better deal is in terms of quantity and the cost.)

Solution Path

The large lemonade is the better buy. Students may use mental math to find that the small and medium drinks hold a total of 23 oz and cost $2.00. The large lemonade is 25¢ less and gives 1 oz more.

Students need to understand that they would be getting more for their money by purchasing the largest one.

2. 2 posters or a puzzle and an art kit

Students may recognize $3.50 as half of $7.00 or they may *guess and check*.

Informal Assessment Ask: What combination of 3 items costs exactly $7? (2 puzzles and a checkers set)

3. The clerk owes Max 4 pennies.

Students may realize that the 2 pennies Max gives the clerk added to 5¢ makes the 7¢ in $7.87. This is the same as if the CD costs $7.85 and Max gives the clerk $10. In either case, the change should be $2.15. They should recognize that $10.02 – $7.87 = $2.15.

4. Pete has 22¢. He can buy any of the combinations shaded in the table. The sale allows him to buy the circled item.

Students may make a table to work out their answer. Encourage more capable students to find all possible answers.

	Popcorn 8¢	Crackers 10¢	Pretzels 13¢	Peanuts 15¢	Almonds 20¢
Popcorn 8¢	16¢ sale 15¢	18¢	21¢	23¢	28¢
Crackers 10¢	18¢	20¢ sale 18¢	23¢	25¢	30¢
Pretzels 13¢	21¢	23¢	26¢ sale 24¢	28¢	33¢
Peanuts 15¢	23¢	25¢	28¢	30¢ sale 25¢	35¢
Almonds 20¢	28¢	30¢	33¢	35¢	40¢ sale 20¢

Name ____________________

4 SHOPPING

Stretch Your Thinking

1. A small lemonade costs $0.75. A medium lemonade costs $1.25. A large one costs $1.75. Which is a better buy, a large lemonade or a small and a medium?

2. Delia chose two items that cost $7.00 altogether. They might be the same or different. What could she be planning to buy?

$1.50 CHECKERS
$2.75 PUZZLE
$4.25 ART KIT
$3.50 POSTER

3. Max buys a CD for $7.87. He gives the clerk a $10 bill and two pennies. The clerk gives Max two dollars, one dime, and one penny. What is wrong?

4. The Snack Palace is having a sale. Pete wants to buy a snack for himself and one for his friend. He has two dimes and two pennies. What can he buy? Name a choice that he could make only because of the sale.

Snack Packs		
Snack	**Cost Each**	**Sale Price**
Popcorn	8¢	2 for 15¢
Crackers	10¢	2 for 18¢
Pretzels	13¢	2 for 24¢
Peanuts	15¢	2 for 25¢
Almonds	20¢	buy 1, get 1 free

4 SHOPPING

Challenge Your Mind

1. Find Out To help students understand the problem, ask:

- What do parts **a**, **b**, and **c** of the problem ask you do? (Find the store that charges the least for one can of beans; find the store that charges the least for two cans of beans; find the store that charges the least for three cans of beans.)
- Explain what "buy 2, get 1 free" means. (You get three cans for $2.00.)
- Which stores have special price deals? (Sonya's and Frederico's)
- What's your estimate? Where you will get each number of cans for the lowest amount? (Answers will vary.)

Solution Paths

a. One can costs $0.67 at Grover's; $0.69 at Sonya's; $1.00 at Frederico's. Comparing these costs shows that Grover's has the lowest price.

b. Two cans cost $1.34 at Grover's ($0.67 + $0.67 = $1.34); they cost $1.33 at Sonya's; they cost $2.00 at Frederico's ($1.00 + $1.00 = $2.00). Comparing these three costs shows that Sonya's has the lowest price.

c. Three cans cost $2.01 at Grover's ($0.67 + $0.67 + $0.67 = $2.01); they cost $2.02 at Sonya's ($1.33 + $0.69 = $2.01). They cost $2.00 at Frederico's (Two cans cost $2.00 and the third is free.) Comparing these three costs shows that Frederico's has the lowest cost.

---

2. yes

Students may *work backward* on this problem. Since Eli chooses the second sale, we know that he paid for only 3 pies. Since he paid $18, we can figure out ($18 ÷ 3 = $6) that each pie costs $6. We can then figure out how much he would have spent had he used the first sale: 4 pies at $6 per pie is $24. Subtract $3 for the coupon and the total is $21. Students may not need to work out the cost with the coupon if they have understood that Eli saved the price of a whole pie ($6) in the sale he chose.

Informal Assessment Ask: In a "buy 3 get 1 free" sale, how can you tell how much you will save? (You will save the cost of one item.) How can you tell if this is a better deal than another? (You have to calculate the costs of the two deals and compare.)

Name ____________________

4 SHOPPING

Challenge Your Mind

1.

a. Where will you pay the least amount for 1 can of beans?

b. Where will you pay the least amount for 2 cans of beans?

c. Where will you pay the least amount for 3 cans of beans?

---

2. Eli is buying pies. There are two sales at the store. If he spends more than $20, he can use a store coupon to take $3 off the total cost. Or he can buy 3 pies and get 1 pie free.

He decides to choose the second sale. Eli pays $18 for 4 pies. Did he get the best deal? Explain your thinking.

4 SHOPPING

Wrap It Up

Book Nook

Part 1 Discuss the problem with the class before they begin to work on their own. Ask:

- How can you use the information in the problem to find an answer to **a**? (Find the total price of the items purchased. Subtract to find the amount of money left. Use a table or chart to figure out the combinations.
- How can you use the information in the problem to find an answer to **b**? (Find the combinations of items that add up to exactly 75¢.)
- How can you use the information in the problem to find an answer to **c**? (Find the cost of the items. Subtract this from the amount of money he pays to find the amount of change he should get. Figure out which coins can be used in the change.)

Solutions

a. Pablo has 80¢ left. He can buy six possible combinations. Accept any correct answer. Possibilities include:

Paperback	0	1	2	3	4	5
Magazines	8	6	5	3	2	0

b. 3 hardcover books or one of 6 different possible combinations. Accept any correct answer. Possibilities include:

Hardcover	0	0	0	1	1	2
Paperback	1	3	5	2	0	1
Magazines	6	3	0	2	5	1

c. 3 coins—1 half-dollar, 2 dimes; \$1.00 – \$0.30 = \$0.70. Use play money to find the fewest coins needed to make \$0.70.

Part 2 Solutions

a. Yes, it changes it. He can also get 10 magazines and 0 paperback books.

b. She can also add 0 hardcover books, 0 paperback books, and 10 magazines to her options.

c. 12 in one of two ways: 1 set of 10 magazines for 75¢ and 2 more at 10¢ or 1 set of 10 magazines for 75¢, 1 more at 10¢, and 1 paperback at 15¢.

Assessment Rubric

3 The student correctly determines the number of each item bought in each case, accurately fulfills all conditions of purchases, understands how the sale prices do and do not supersede the original prices, and supplies all possible answers.

2 The student correctly determines the number of each item bought in each case, accurately fulfills all conditions of purchases, has some difficulty understanding how the sale prices do and do not supersede the original prices, and supplies at least one of the possible answers.

1 The student correctly determines the number of each item bought in most cases, accurately fulfills most conditions of purchases, but has difficulty understanding how the sale prices supersede the original prices and does not supply all the possible answers.

0 The student can determine the number of each item bought in a limited number of cases, but lacks understanding of how the sale prices supersede the original prices and does not supply the correct answers.

Name ____________________

4 SHOPPING

Wrap It Up

Book Nook

Part 1

The library is having a sale. Look at the prices on the table.

a. Pablo has $1.55. He wants to buy 3 hardcover books. How many paperback books and magazines can he buy?

b. Sue has 75¢. What can she buy and have no change left over?

c. Bob buys 2 paperback books. If he gives the librarian $1, what is the least number of coins he could get in change? What are they?

Part 2

The librarian decides to make some special deals. He puts up this sign next to the other signs.

a. Could this change what Pablo buys? If so, how?

b. Could this change what Sue buys? If so, how?

c. What is the greatest number of books and magazines Bob could buy for $1?

5 BUDGETS
Introduction

This section encourages students to think about making choices regarding money—setting goals, making budgets, and spending wisely.

Understanding Budget Problems

As students prepare to solve budgeting problems, they will need to determine what moneys are available and what needs to be earned or saved for a particular purpose. Consider the following problems:

Wanda got $53 in gifts for her birthday. She already had saved $18 from her allowance. If she saves $2 a week, for how long will she have to save to buy a keyboard that costs $119?
To solve this problem, students need to:

- Total the amount that Wanda already has.
- Find the difference between what she has and the final cost.
- Figure out how many weeks it will take her to save this amount.
- Understand that this is a multistep problem involving several steps that must be done consecutively.

Joel wants to buy two different kinds of cereal. He has $6.00. Which cereals should he buy to have the most money left over? How much money will be left?

Wheat cereal	$3.89
Oat cereal	$2.26
Bran cereal	$3.38
Corn cereal	$2.59

To solve this problem, students need to:

- Determine the pair of cereals that costs the least.
- Subtract the total cost from $6.00.

As students work with budgeting problems, ask these key questions as needed:

- What amounts of money do you see in the problem?
- How do the different amounts of money relate to each other?
- Do any choices have to be made to solve the problem?

Solving Budget Problems

Your students will be more successful if they can combine general problem-solving strategies with methods specific to budget problems such as *use or make a table* and *guess and check*. They may find using play money useful. Refer to page *vi* for a discussion of strategies.

Solving a Budget Problem by Making a Table
Making a table to view the results of different combinations helps keep the information orderly and easy to analyze. Some students, especially the younger ones, may not be ready to make a table but can complete or use a table that you set up.

Example: Jed has $2.00. What two items can he buy?

Baseball cards	$0.75 a pack
Marbles	$0.89 a bag
Jacks	$0.68 a bag
Pick-up-sticks	$1.19 a carton
Card games	$0.93 a pack

The table below gives the total cost of each pair of items and shows whether or not Jed can buy them. Students can complete the entire table or just enter the *yes/no* decisions. Note: The shaded boxes are duplicates of other boxes.

	Baseball cards $0.75	**Marbles $0.89**	**Jacks $0.68**	**Pick-up sticks $1.19**	**Card games $0.93**
Baseball cards $0.75	$1.50 yes	$1.64 yes	$1.43 yes	$1.94 yes	$1.68 yes
Marbles $0.89	$1.64 yes	$1.78 yes	$1.57 yes	$2.08 no	$1.82 yes
Jacks $0.68	$1.43 yes	$1.57 yes	$1.36 yes	$1.87 yes	$1.61 yes
Pick-up sticks $1.19	$1.94 yes	$2.08 no	$1.87 yes	$2.38 no	$2.12 no
Card games $0.93	$1.68 yes	$1.82 yes	$1.61 yes	$2.12 no	$1.86 yes

Assessment

✓ **Informal Assessment** A suggestion for informal assessment will be found on each Try It Out, Stretch Your Thinking, and Challenge Your Mind page. The recommended question will help focus students' attention on one part of the problem-solving process.

Assessment Rubric An assessment rubric is provided for each Wrap It Up. Students' completed work may be added to their math portfolios.

Thinking About Budgets

These problems will help students use their number sense and estimation skills as preparation for solving budget problems. Present one problem a day as a warm-up. You may choose to read the daily problem aloud, write it on the board, or create a transparency.

1. **Matt saves $0.30 from his allowance every other week. He has been saving for 10 weeks. Does he have more than $1.75? Explain.**
(No; 10 weeks means he has saved $0.30 five times; $5 \times \$0.30 = 1.50, \$1.50 < \$1.75$.)

2. **Julio earns $2.50 every week for baby-sitting. He is saving to buy a guitar. So far he has saved $27.50. Has he been saving for more than 10 weeks?**
(Yes; mentally compute that at 10 weeks he would have $10 \times \$2.50$ or $25.00.)

3. **Goldfish cost $0.80 each. Next week, if you buy one you get one free. Rosa wants to get 8 fish for her tank. Will she save more than $3.00 if she buys them next week? Explain.**
(Yes; mentally figure that Rosa saves the price of 4 fish; $\$0.80 \times 4 = \$3.20, \$3.20 > \$3.00$.)

4. **Tina wants to buy a sweatshirt for $15 and a radio for $26. Every week she saves $1 for the sweatshirt and $2 for the radio. Which item will she be able to buy first?**
(The radio; the sweatshirt is $15 and it will take 15 weeks to save the money, the radio is $26 and it will take 13 weeks to save the money, $13 < 15$.)

5. **Jimmy wants to buy a packet of tomato seeds for $1.23, a trowel for $3.99, and a watering can for $5.02. Will $10.00 be enough? Explain.**
(No; estimate that $\$3.99 + \$5.02$ is over $9.00. $\$9.00 + \$1.23 > \$10.00$.)

5 BUDGETS

Solving Budget Problems

This lesson will help students focus on the kinds of questions they need to ask in order to solve problems involving budgets. Emphasize that there may be several approaches to solving a problem.

Using the Four-Step Method

Find Out

- ○ The problem asks students to figure out a possible combination of items that Marty can buy. Encourage students to describe the problem in their own words so they understand that there are several steps involved in discovering what the possibilities might be.
- ● Remove the bus fare from the amount Marty has. Estimate how much he can spend.
- ○ Estimate Marty has $2.00 – $0.75 or $1.25 to spend. The first drink and snack (apple juice and banana) would cost $1.00 altogether. Choose other combinations with similar or slightly greater costs.

Make a Plan

- ● The difference between $2.04 and $0.75 is the exact amount that Marty has to spend.

Solve It

- ○ banana and any drink; apple and any drink; trail mix and apple juice; yogurt and apple juice

Here are two possible solution paths:

Make a Table Students may make a table to find all the pairs with a purchase price of less than $1.29.

	Apple Juice ($0.50)	Cranberry Juice ($0.75)	Milk ($0.65)	Orange Juice ($0.70)
Banana ($0.50)	$1.00	$1.25	$1.15	$1.20
Apple ($0.45)	$0.95	$1.20	$1.10	$1.15
Trail Mix ($0.75)	$1.25	$1.50	$1.40	$1.45
Yogurt ($0.65)	$1.15	$1.40	$1.30	$1.35

Use Logical Reasoning Students may use mental math to figure that a banana and apple juice will total exactly $1.00; and *use logical reasoning* to conclude that this is less that $1.25.

Look Back

- ● Have students compare their results with their estimates. Ask: Is your answer reasonable when you compare it with your estimate? How did estimating help you solve the problem?
- ○ A way to check the solution is to solve by a different method. Encourage students to share their solution methods.

Name ____________________

5 BUDGETS

Solving Budget Problems

Marty has $2.04. He needs to keep $0.75 for the bus ride home. He wants to get a drink and a snack at the snack bar. What can he buy?

Drinks		Snacks	
Apple Juice	50¢	Banana	50¢
Cranberry Juice	75¢	Apple	45¢
Milk	65¢	Trail Mix	75¢
Orange Juice	70¢	Yogurt	65¢

Find Out	○ What question must you answer to solve the problem? • Do you need to find the most money Marty can spend on a drink and a snack to solve the problem? Explain. ○ What is a reasonable estimate?
Make a Plan	• How would subtracting $0.75 from $2.04 help you find the answer?
Solve It	○ Find the answer. Use your plan. Keep a record of your work.
Look Back	• How does your answer compare with your estimate? ○ How would you check to see whether your answer is correct?

5 BUDGETS

Try It Out

1. Find Out To help students understand the problem, ask:

- What does the problem ask you to find? (what Sandy is able to buy besides the required items)
- Would it help you to know:

a. How much money the required items cost altogether? (yes)

b. How much money she has left after paying for the required items? (yes)

- What is your estimate of what Sandy can buy? (Students may estimate that she has $50 and must spend about $30. Based on this, they may guess that she can get at least one of each kind of photo, and probably two of at least one kind of photo.)

Solution Path

1 team photo and 2 player photos or 2 team photos or 5 player photos (or any combination with less than those amounts, such as 4 player photos, 3 player photos, and so on)

$18.50 + $12.25 = $30.75,
$47.00 – $30.75 = $16.25,
$7.75 + $3.25 + $3.25 = $14.25,
$7.75 + $7.75 = $15.50,
5 × $3.25 = $16.25

2. 6 beads; 5 × $0.30 = $1.50; 6 × $0.25 = $1.50

Students may mentally calculate this answer.

3. 10 weeks; $1.79 + $5.96 + $1.50 = $9.25

Some students may think that Jules will have saved enough in 9 weeks. Point out that at the end of 9 weeks he will have saved $9. He needs the tenth week's savings to get the last $0.25.

4. $6.40; 16 × $0.40 = $6.40

5. 12 weeks; $0.80 × 10 = $8.00, $0.80 × 1 = $0.80,
$8.00 + $0.80 = $8.80 (11 weeks),
$8.88 – $8.80 = $0.08

One more week is needed to get the last $0.08.

Informal Assessment Ask: If Huan saved only $0.40 a week, how long would he have to wait? (23 weeks)

6. $14.25; $0.75 + $0.75 + $0.75 +
$5.00 + $3.50 + $3.50 = $14.25

7. 11 weeks; 10 × $1.25 = $12.50,
$12.50 + $1.25 = $13.75

Students may also *work backward* to find the answer.

Name ____________________

5 BUDGETS

Try It Out

1. Sandy has $47 for baseball this summer. She has to pay for sign-up and a shirt and cap. What else can she fit in her budget?

Sign-up fee	$18.50
Shirt and cap	$12.25
Optional Team Photo	$7.75 each
Optional Player Photo	$3.25 each

2. Midori puts 30¢ in her piggy bank every week. She wants to buy some glass beads that cost 25¢ each. How many beads can she buy after 5 weeks?

3. Jules wants to get some birthday presents for his little brother. He chooses a toy car for $1.79, a game for $5.96, and a poster for $1.50. Jules saves $1 a week. How many weeks will it take him to save enough money?

4. Winona saves $0.40 every week from her allowance. She has been saving for 16 weeks. How much has she saved?

5. Huan saves $0.80 a week from his allowance to put toward buying a model car that costs $8.88. How many weeks will it be before he can buy it?

6. Alice, John, and Jeannie are planning a party for 12. They want to buy matching napkins, plates, and cups. Look at the prices below. How much do they need to save?

Napkins, Package of 4	$0.75
Plates, Package of 12	$5.00
Cups, Package of 6	$3.50

7. Ella earns $1.25 every week for helping her neighbor with his garden. She saves it all, and now she has $13.75. For how many weeks has she been saving?

5 BUDGETS

Stretch Your Thinking

1. **Find Out** To help students understand the problem, ask:

- What question does the problem ask you to answer? (What combination of instruments could the class buy and stay within their budget?)

Make a Plan To help students decide on a strategy to follow, ask:

- Why can't the class buy two recorders? (They must buy two different kinds of instruments, not two of one kind; note, the pair of maracas is considered one instrument.)
- Is this a problem that has one answer or more than one? (More than one possible answer.)

Solution Paths

Students may *work backward*, *use objects*, or *make a table*. This table shows the total cost of each possible combination.

	Small drum $4.35	**Tambourine $5.60**	**Recorder $3.99**	**Triangle $2.78**	**Maracas $6.48**
Small drum $4.35	X	$9.95 yes	$8.34 yes	$7.13 yes	$10.83 no
Tambourine $5.60	$9.95 yes	X	$9.59 yes	$8.38 yes	$12.08 no
Recorder $3.99	$8.34 yes	$9.59 yes	X	$6.77 yes	$10.47 no
Triangle $2.78	$7.13 yes	$8.38 yes	$6.77 yes	X	$9.26 yes
Maracas $6.48	$10.83 no	$12.08 no	$10.47 no	$9.26 yes	X

2. 14 weeks; $14 \times \$1.25 = \$17.50, \$17.50 > \$17.00$

Students can *guess and check* that the number of weeks is greater than 10 and less than 16: mentally calculate that $10 \times \$1.25 = \$12.50$ and $16 \times \$1.25 = \$20.00; \$20 > \$17$.

Ask: Will Joe have extra money when he has enough to buy the hat? (yes)

3. $5.85; $\frac{1}{2} \times \$0.50 = \$0.25, 9 \times \$0.25 = \$2.25$, $\frac{1}{2} \times \$0.80 = \$0.40, 9 \times \$0.40 = \$3.60$, $\$2.25 + \$3.60 = \$5.85$

✓ **Informal Assessment** Ask: What steps are required to solve this problem? (Find out how much allowance Ronald is saving per week in each case; then determine the amount of money he saved in each set of 9 weeks and add the two amounts together.)

4. 5 weeks; Jennie will be ready with her share in 4 weeks. Estimate the cost of the computer game as $20. Each child is supposed to pay half, about $10; $\$2 \times 5 = \$10$; $\$2.75 \times 4 = \$11$.

5. Magazine subscription; use rounding, estimating, and adding on to determine that yo-yo = 16 weeks, magazine = 13 weeks, and in-line skates = 19 weeks. For the magazine subscription, $\$0.85 \times 10 = \$8.50$; $\$8.50 + 3 \times \$0.85 = \$11.05$.

Name ______________________

5 BUDGETS

Stretch Your Thinking

1. Penelope's class wants to buy two different kinds of instruments for their school play. They have $10 to spend. What instruments can they buy?

	small drum	$4.35
	tambourine	$5.60
	recorder	$3.99
	triangle	$2.78
	pair of maracas	$6.48

2. Joe always gets a baseball cap for his favorite team in the World Series. Team hats cost $17. Joe's allowance is $1.25 a week. How quickly can he save the money to buy the hat?

3. Ronald saves half his weekly allowance of 50¢ to buy books. After 9 weeks, his parents raise his allowance to 80¢ a week. If Ronald saves half his allowance for another 9 weeks, how much money will he have to spend on books?

4. Jennie and Eddie each get a weekly allowance of $2.75. Each will pay half the cost of a computer game that sells for $19.95. Jennie saves all her allowance. Eddie spends his allowance, but he saves $2.00 every week from washing cars. How many weeks will it be before they can buy the game?

5. Every week, Denise saves $0.65 of her allowance for a yo-yo, $0.85 for a magazine subscription, and $2.00 for in-line skates. The yo-yo costs $9.95. The magazine subscription costs $10.50. The in-line skates cost $38.00. Which item will Denise be able to buy first?

5 BUDGETS

Challenge Your Mind

1. Find Out To help students understand the problem, ask:

- What baked goods could the class choose from? (brownies, muffins, cookies, bagels, rolls)
- What are the limits for each of the lettered questions? (**a.** \$10 to spend on identical items; **b.** \$10 to spend on assortment of items; **c.** Spend as much of the \$10 as possible)
- For this problem, you may want to give the definition of *assortment* as including a minimum of 3 each of 4 different items.

Solution Paths

a. cookies or rolls; cookies: \$0.75 × 13 = \$9.75; rolls: \$0.50 × 13 = \$6.50. Thirteen of any of the other baked goods total more than \$10.

b. Possible answers:
3 brownies, 3 muffins, 3 cookies, 4 rolls;
3 brownies, 3 bagels, 3 cookies, 4 rolls;
3 muffins, 3 bagels, 4 cookies, 3 rolls;
3 muffins, 3 bagels, 3 cookies, 4 rolls.
Students may *make a table* or *act out or use objects.* Any assortment of 3 each of 4 or more different types of baked goods for a total of 13 items costing under \$10 is acceptable.

c. 7 brownies, 6 rolls; Students can use the table below as a starting point. The question then becomes whether they can spend exactly \$10.

Brownies \$1.00	Muffins \$0.90	Bagels \$0.85	Cookies \$0.75	Rolls \$0.50	Total Cost
3	3	0	3	4	\$9.95
3	0	3	3	4	\$9.80
0	3	3	4	3	\$9.75
0	3	3	3	4	\$9.50
3	3	0	3	4	\$9.95
7	0	0	0	6	\$10.00

Informal Assessment Ask: If the class had \$13 to spend, how might the choices be different? (Answers will vary but may include: They could spend \$1 per person or they could give each worker a brownie or two rolls. They might give each one roll and some baked goods to split and share.)

---

2. yes; \$4.97 + \$1.42 = \$6.39,
\$0.15 + \$0.09 + \$0.18 + \$0.13 + \$0.14 + \$0.21 + \$0.17 = \$1.07,
\$1.07 × 6 = \$6.42; \$6.42 > \$6.39

Name ____________________

5 BUDGETS

Challenge Your Mind

1. Students in Mr. Ingalls' class want to buy a treat for each of the 13 people who work in the school kitchen. The class has $10. How might the students spend their money if:

a. they want to give every worker the same treat?

b. they want to get an assortment to make it easier for the workers to find something they like?

c. they want to spend as much of the $10 as possible?

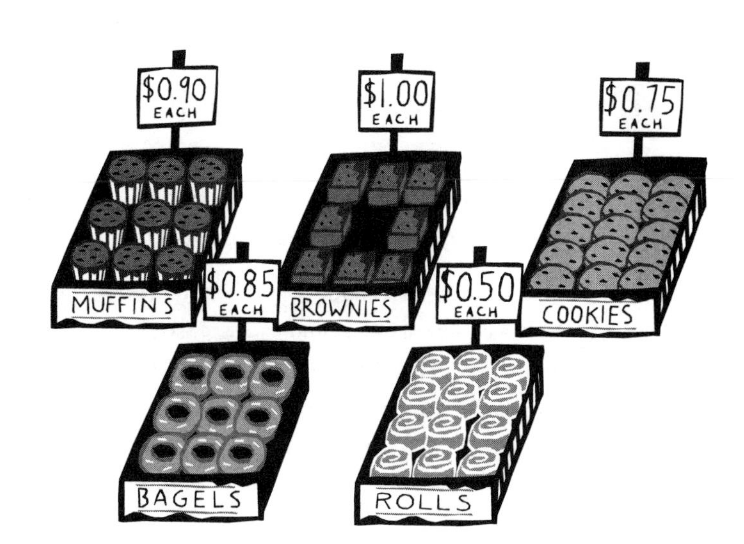

Weekly Amounts from Molly's Friends	
Jilly	$0.15
Christopher	$0.09
Abigail	$0.18
Ricardo	$0.13
Sara	$0.14
Ben	$0.21
Megan	$0.17

2. Molly's birthday is in 6 weeks. Her friends want to buy her a present. Jilly saw a necklace that costs $4.97. Sara wants to decorate the gift box with a small plastic animal that costs $1.42. Each of Molly's friends is going to give some of his or her allowance. Can they save enough to buy the gifts before Molly's birthday?

5 BUDGETS

Wrap It Up

Home Sweet Terrarium

You might wish to discuss the problem with the class before they begin to work on their own. Ask:

- Why would it make sense to figure out the cost of 2 bags of the cheapest gravel and 1 plant first? (They are necessary items. Once you have figured them out, you know how much money you have left to use as you wish.)

Solutions

a. If the children buy 2 bags of black or brown gravel and 1 plant, they can get any combination of 2 animals. If they buy 2 bags of black or brown and 2 plants, they can get any combination of 2 animals. With 2 bags of colored gravel and 1 plant, they can get any combination of 2 animals. With 2 bags of colored gravel and 2 plants, they can get any combination of 2 animals except 2 fire-belly toads.

b. If they buy 2 bags of black or brown gravel and 1 plant, they can get 5 fire-bellied newts or 4 fire-bellied newts and 1 red-spotted newt, for a total of 5 animals.

c. With 4 bags of colored gravel and 2 plants, they can only get either a fire-bellied or a red-spotted newt.

Possible Solution Paths

a. $3 \times \$2.99 = \$8.97$
$\$25.00 - \$8.97 = \$16.03$
$\$5.99 + \$4.99 = \$10.98$
$\$10.98 + \$8.97 = \$19.95$

b. $2 \times \$2.99 = \$5.80$
$\$5.80 + \$2.99 = \$8.97$
$\$25.00 - \$8.97 = \$16.03$

Round cost of least-expensive animal, the fire-bellied newt, to \$3.00; count up to \$16.00.

c. $\$15.96 + \$5.00 = \$20.96$
$\$25.00 - \$20.96 = \$4.04$

Assessment Rubric

3 The student writes problems that meet all the criteria and finds the correct solutions for them.

2 The student writes problems that meet some of the criteria and finds solutions for them.

1 The student writes problems but does not meet the criteria given and/or does not find the correct solutions.

0 The student neither writes problems that meet the criteria nor finds the correct solutions.

Name ______________________

5 BUDGETS

Wrap It Up

Home Sweet Terrarium

Fire-belly Toad	$6.25
Green Tree Frog	$4.99
Gray Tree Frog	$4.50
Red-spotted Newt	$3.75
Fire-bellied Newt	$2.99

Plants	$2.99 or 2 for $5.00
Black or Brown Gravel	$2.99
Colored Gravel (red, blue, yellow, green)	$3.99

The Bunduki children are planning a terrarium. They can spend $25.00. They already have everything they need except animals, gravel, and plants. The children must buy 2 bags of gravel and 1 or 2 plants.

a. What is one set of things the children can buy that includes 2 animals?

b. What is one set of things that has the greatest possible number of animals?

c. If the children buy 4 bags of colored gravel and 2 plants, what animals could they buy?

6 MONEY PUZZLES

Introduction

This section focuses on money puzzles and problems with an extra mental challenge. These puzzles may require an unusual strategy or approach to solve them.

Understanding Money Puzzles

As students prepare to solve money puzzles, they may need to consider decoding, working within limitations, finding a pattern, or creating a chain of logical reasoning. Consider the following problems:

Say A = 1¢, B = 2¢, C = 3¢, . . . Z = 26¢. You have the following letters: K, C, Q, L, M, I, U. What is the most expensive word you can make?
To solve this problem, students will need to:

- Recognize the code pattern.
- Apply a value to the given letters.
- Choose a method for determining the most valuable word.
- Check that their results seem reasonable.

The Acme Book Company charges $0.50 to ship one book, $0.75 to ship two books, $1.00 to ship three, and $1.25 to ship four. How much would it cost to ship six books?
In a problem such as this, students must:

- Recognize and analyze the pattern.
- Extend the pattern to find the solution.
- Understand that this is a multistep problem—they must first find the pattern for the prices given and then extend the pattern to find the answer.

Ann has $0.25 more than Claire. Together they have $1.25. How much money does each of them have?
In a problem such as this, students must:

- Understand that this is a multistep problem. Students might use *guess and check* or they might *work backward.* First they would subtract the difference, $1.25 – $0.25 = $1.00. Then they split the $1.00 into $0.50 and $0.50. Finally they add the $0.25 difference to one of the $0.50.

As students work with money puzzles, use these prompts as needed:

- Describe the puzzle in your own words.
- Does it have multiple steps?
- Could there be more than one answer?
- What methods might be used to solve it?

Solving Money Puzzles

Your students will be more successful if they can apply logical reasoning and use general problem-solving strategies such as *make a table, look for a pattern, make it simpler, work backward, guess and check,* and *brainstorm.* Refer to page *vi* for a discussion of the ten solution strategies.

Solving a Money Puzzle Using a Table In puzzles, it is particularly useful to *use a table* to record input in conjunction with another strategy.

Example: George gets twice as much allowance as his younger sister Sally. Together, they get $9. How much do they each get?

George's Allowance	Sally's Allowance	Total	
$8	$1	$9	no
$7	$2	$9	no
$6	$3	$9	yes

Solving a Money Puzzle Using Logical Reasoning and an Organized List

Example: You have 5 coins. You can use these 5 coins to make an exchange with the least number of coins for a dime, then for a quarter, and then for a half-dollar. What 5 coins do you have?

Set up a list:

Coin	Coins for exchange
Dime	2 nickels
Quarter	2 dimes, 1 nickel
Half-dollar	1 quarter, 2 dimes, 1 nickel

The 5 coins used for exchanges are 1 quarter, 2 dimes, and 2 nickels.

Assessment

Informal Assessment A suggestion for informal assessment will be found on each Try It Out, Stretch Your Thinking, and Challenge Your Mind page. The recommended question will help focus students' attention on one part of the problem-solving process.

Assessment Rubric An assessment rubric is provided for each Wrap It Up. Students' completed work may be added to their math portfolios.

Thinking About Money Puzzles

These problems will help students use their number sense and estimation skills as preparation for solving money puzzles. Present one problem a day as a warm-up. You may choose to read the daily problem aloud, write it on the board, or create a transparency.

1. **Five coins are worth 51¢. Can more than one coin be a penny? Explain.**
   (No; If there were more pennies, there would have to be 6 pennies and that would make more than 5 coins total.)

2. **Vowels are worth 4¢. All other letters are worth 2¢. Which name is worth more, Aaron or Susan?**
   (Aaron; The words have the same number of letters, but Aaron has one more vowel.)

3. **Misha sells 1 balloon for $1, 2 balloons for $2, 4 balloons for $3, and 8 balloons for $4. Would 30 balloons cost more than $6?**
   (No; following the pattern, the number of balloons is doubled for every additional dollar, so for $5 you get 8 balloons doubled, or 16 balloons, and for $6 you get 16 balloons doubled, or 32 balloons; 32 > 30.)

4. **Suppose you have 17 coins. None of them are dollar coins. Could they be worth more than $9.00? Why or why not?**
   (No; the largest coins you could have would be half-dollars and if all of the coins were half-dollars they would be worth $8.50.)

5. **Say A = 1¢, B = 2¢, C = 3¢, . . . Z = 26¢. You need to form a three-letter word that is worth less than 35¢. Can you start your word with the letter Q? Explain your thinking.**
   (No; Q would have to be followed by U and the total for just those two letters is 38¢.)

6 MONEY PUZZLES

Solving Money Puzzles

This lesson will help students focus on the kinds of questions they need to ask themselves in order to solve money puzzles. Emphasize that there may be several approaches to solving a problem.

Using the Four-Step Method

Find Out

- ○ The problem asks students to find the values of the 13 unidentified coins that Charles has. Encourage students to describe the problem in their own words so that they understand how to use the given information.
- ● 15 coins
- ○ $3.99; All 15 coins are worth less than $5.00. Two of the coins are half-dollars, so the remaining coins are worth less than $5.00 – $1.00, or $4.00. The greatest amount less than $4.00 is $3.99.
- ● $3.01; All 15 coins are worth more than $4.00. We know that two of the coins total $1.00. The remaining 13 coins must be worth more than $4.00 – $1.00, or $3.00. The least amount greater than $3.00 is $3.01.
- ○ There must be at least 12 quarters. The least the 13 coins can total is $3.01 and you need 12 quarters to get over $3.00.

Make a Plan

- ● Students will most likely *make a table, work backward, use objects,* or *guess and check.*

Solve It

- ○ There are three possible answers. Using the *make a table* strategy:

Half-Dollar	Quarter	Dime	Nickel	Total Value
2	13	0	0	$4.25
2	12	1	0	$4.10
2	12	0	1	$4.05

Look Back

- ● Have students compare their results with their estimates. Ask: Is your solution reasonable when compared with your estimate? How did estimating help you solve the problem?
- ○ A way to check the solution is to solve by a different method. Encourage students to share their solution methods in small groups.

Name ______________________

6 MONEY PUZZLES

Solving Money Puzzles

Charles has 15 coins. Two coins are half-dollars. The other 13 coins could be quarters, dimes, and nickels. Together the coins are worth more than $4 and less than $5. What coins might he have?

Find Out	○ What question must you answer to solve the problem? • How many coins does Charles have in all? ○ What is the most the 13 mystery coins could be worth according to the problem? • What is the least they could be worth? ○ What is a reasonable estimate?
Make a Plan	• How can you find the possible combinations of coins that Charles might have?
Solve It	○ Solve the problem using your plan. Keep a record of your work.
Look Back	• How does your answer compare with your estimate? ○ How would you check to see whether your answer is correct?

6 MONEY PUZZLES

Try It Out

1. Make a Plan To help students think about working toward an answer, ask:

- Does this problem have one answer or more than one answer? How do you know? (It has more than one answer, because there are so many foods.)
- How could making a list of all the letters and their values help you? (You would not have to keep counting the values each time you used a letter, and you could estimate the value of a food without doing the addition.)
- What kinds of words will give low values? high values? (short words and words with letters from the beginning of the alphabet; long words and words with letters from the ends of the alphabet)
- What are the names of some foods you could try out? (apple, orange, yogurt, bread, chicken)

Solution Paths

A	1¢	J	10¢	S	19¢
B	2¢	K	11¢	T	20¢
C	3¢	L	12¢	U	21¢
D	4¢	M	13¢	V	22¢
E	5¢	N	14¢	W	23¢
F	6¢	O	15¢	X	24¢
G	7¢	P	16¢	Y	25¢
H	8¢	Q	17¢	Z	26¢
I	9¢	R	18¢		

Answers may vary. Some possible answers:

a. pea = 22¢, cake = 20¢

b. peanut butter = $1.63

c. apple = 50¢; orange = 60¢

2. 1 quarter and 3 dimes; 25¢ + 10¢ + 10¢ + 10¢ = 55¢

Students may *make a table* or *guess and check*. Some students may *use logical reasoning* to figure that there must be at least 1 quarter (4 dimes is not enough) and mental math to figure that the remaining 30¢ is equivalent to 3 dimes.

3. a. 1 dime, 1 nickel, 5 pennies;
10¢ + 5¢ + 1¢ + 1¢ + 1¢ + 1¢ + 1¢ = 20¢

b. 1 dime, 2 nickels; 10¢ + 2(5¢) = 20¢

Ask: What coins are the equivalent of 5 other coins? (A nickel equals 5 pennies, a quarter equals 5 nickels, and a half-dollar equals 5 dimes.)

4. Louisa has $1.25, Arnold has $1.50;
$1.25 + $1.50 = $2.75

5. There are three possible solutions:

$$\begin{array}{r}432\\+432\\\hline 864\end{array}\qquad\begin{array}{r}427\\+427\\\hline 854\end{array}\qquad\begin{array}{r}417\\+417\\\hline 834\end{array}$$

Students may use *guess and check* to solve. Most students should find the first solution because it doesn't require any regrouping.

6. Answers will vary. Possible answers include: store (21¢); shred (31¢); hard (27¢); noted (21¢); sorted (30¢)

Informal Assessment Ask: Make a word with a value under 20¢. What letters will you try to use? Why? (Answers will vary.)

Name ____________________

6 MONEY PUZZLES

Try It Out

1. Let A = 1¢, B = 2¢, C = 3¢, and so on up to Z = 26¢. Think about foods you like. Figure out the value of their names. Name a food you can buy for each amount of money.

a. Less than a quarter

b. More than a dollar

c. Exactly 50¢ or exactly 60¢

2. There are four coins in the bank. Their total value is 55¢. What are they?

3. Dominic has 7 coins worth 20¢. At the store, he trades 5 coins for 1 coin. He still has 20¢.

a. What coins did he start with?

b. What coins does he have now?

4. Louisa and Arnold each have some coins. Together they have $2.75. If Louisa gets 25¢ more, they will have the same amount of money. How much does each have now?

5. Each letter below is worth 1¢, 2¢, 3¢, 4¢, 5¢, 6¢, 7¢, or 8¢. No two letters are worth the same amount. The letter O is worth 4¢. How much are the letters E, T, W, and N worth?

```
  O N E
+ O N E
-------
  T W O
```

6. Use these letter tiles to make 5 different words. Use each tile only once in a word. Each word must be worth more than 20¢ and less than 40¢.

6 MONEY PUZZLES

Stretch Your Thinking

1. Find Out To help students understand the problem, ask:

- What question does the problem ask you to answer? (how to use letters of given value to make the word of the greatest value and of the least value)
- Is this a problem that has one answer or more than one possible answer? (more than one possible answer)

Solution Paths

a. Possible answers: dreams 38¢, medals 40¢, marred 40¢, snarled 42¢, ladled 42¢, saddled 47¢, slammed 51¢

Students may realize that endings like –*s* and –*ed* can be added to increase word value.

b. Possible response: a 3¢

2. Any multiple of 2 and 5 works. (10¢, 20¢, 30¢, 40¢, 50¢, 60¢, 70¢, 80¢, 90¢)

Ask: What digit is in the ones place in all numbers that are multiples of ten? (zero) How does this help you? (It makes it easier to compute mentally.)

3. a. 11¢; the first nail costs 1¢, the second nail costs 2¢ (add 1¢); the third nail costs 4¢ (add 2¢), the fourth nail costs 7¢ (add 3¢), the fifth nail costs 11¢ (add 4¢).

b. 92¢; the sixth nail will cost 16¢ (add 5¢), the seventh will cost 22¢ (add 6¢) and the eighth will cost 29¢ (add 7¢);
29¢ + 22¢ + 16¢ + 11¢ + 7¢ + 4¢ + 2¢ + 1¢ = 92¢.

✓ **Informal Assessment** Ask: Would it be smarter to buy two horseshoes at the same time or separately? Why? (Separately, because 16 nails cost more than two purchases of 8 nails.)

4. There are three possible combinations:
2 nickels, 3 dimes, 3 quarters ($1.15);
2 nickels, 2 dimes, 4 quarters ($1.30);
3 nickels, 2 dimes, 3 quarters ($1.10).

5. B = 6¢, A = 7¢, Y = 4¢, T = 1¢,
W = 3¢, I = 5¢, N = 2¢, S = 8¢

$$\begin{array}{r} 6{,}764 \\ +\,6{,}764 \\ \hline 13{,}528 \end{array}$$

Students may use *guess and check* or *use logical reasoning* to determine the following:

B = 6¢. So, B + B = 12¢. Therefore T = 1¢. Also, Y cannot be 2 because either W or N must be 2. Y cannot be 3 because S is not 6. If Y is 4, S is 8. Then N is 2. Since B + B in the thousands place is not N, A + A must be greater than 8. A cannot be 5 because I is not 1. A must be 7. That means I is 5 and W is 3.

Name ______________________

6 MONEY PUZZLES

Stretch Your Thinking

1. Use these letter tiles to make words. You can use each letter as many times as you wish in each word.

a. What is the greatest value word you can make?

b. What word can you make that is worth the least?

---

2. Find 3 amounts of money less than \$1.00 that can be made with just dimes *and* with just nickels.

3. A blacksmith charges by the nail for horseshoes. He charges 1¢ for the first nail, 2¢ for the second, 4¢ for the third, 7¢ for the fourth and so on.

a. What amount does he charge for the fifth nail? Explain.

b. If a horseshoe requires 8 nails, what would it cost?

---

4. Toby has 8 coins. He has only nickels, dimes, and quarters. He has at least 2 of each kind of coin. More than 2 of his coins are quarters. Less than one half of his coins are dimes.

a. What coins could Toby have?

b. How much are they worth?

5. Each letter is worth either 1¢, 2¢, 3¢, 4¢, 5¢, 6¢, 7¢, or 8¢. Different letters are worth different amounts. Figure out how much each letter is worth. Hint: B = 6¢.

```
   B A B Y
 + B A B Y
 ---------
 T W I N S
```

6 MONEY PUZZLES

Challenge Your Mind

1. Find Out To help students understand the problem, ask:

- What are the values that shapes might have? (5¢, 10¢, 15¢, 20¢, 25¢, 30¢, 35¢)

Solution Paths

a. 75¢;

2 squares	30¢
1 circle	5¢
1 triangle	10¢
1 hexagon	+ 30¢
	75¢

b. Answers will vary, but shapes should add up to $1.50.

c. Answers will vary, but shapes should add up to $2.00.

Note that some drawings may be counted in more than one way. For example, a student making a hexagon with three isosceles triangles added to form a larger triangle (one on either side of the base, and one at the top) may or may not count the larger triangle (perhaps depending on whether or not it was deliberately drawn). If such a case occurs, draw students' attention to the possibilities and accept all reasonable responses.

2. There are six possible combinations.

Pennies	Nickels	Dimes	Total Value
2	5	5	$0.77
2	4	6	$0.82
2	3	7	$0.87
2	2	8	$0.92
2	1	9	$0.97
2	0	10	$1.02

Informal Assessment Ask: How could this problem be made easier by changing or adding rules? How could it be made more difficult? (Students may suggest that the problem would be easier if any coin could be used and/or any number of each coin. Students may suggest that the problem would be more difficult if more rules were added.)

Name ____________________

6 MONEY PUZZLES

Challenge Your Mind

1. Use the table below.

Shape	Value
Circle	5¢
Triangle	10¢
Square	15¢
Rectangle (not a square)	20¢
Pentagon	25¢
Hexagon	30¢
Octagon	35¢

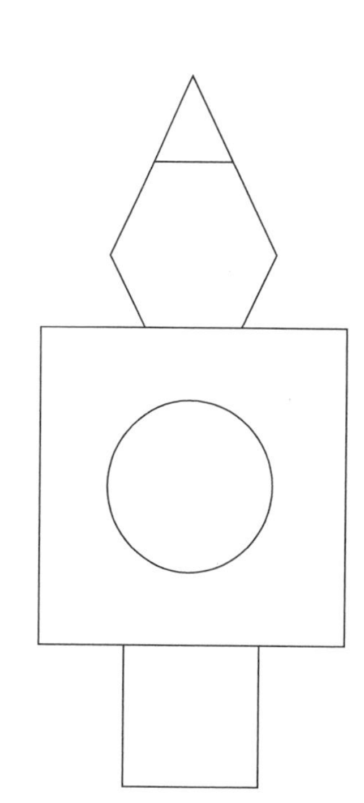

Show all your work.

a. How much is this picture worth?

b. Make a drawing worth twice as much.

c. Make a drawing that is worth $2.00.

---

2. Jean has 12 coins.

- She has only pennies, nickels, and dimes.
- She has more than 4 dimes.
- Fewer than one half of her coins are nickels.
- She has 2 pennies.

What coins could Jean have?
How much are they worth?

6 MONEY PUZZLES

Wrap It Up

What's in a Name?

Part 1 You might wish to discuss problem-solving strategies with the class before they begin to work on their own.

- A few pages from an old telephone book would facilitate students' problem-solving and creation of names in this section.
- Students might use the following process as they create problems:

  Ask students to work in small groups. Ask groups to make suggestions about how to word problems and arrive at solutions clearly.

  Tell students to write each problem on the front of a 5 × 8 card making the language as clear as possible. On the back of the card, have them write the solution.

Solutions

a. Answers will vary depending on students' names.

b. Suzanne; Steven

c. Answers will vary.

Possible Solution Paths

b. Millicent = 97¢; Suzanne = $1.00; Rosalie = 79¢; Steven = 85¢; Zachary = 82¢; Mitchell = 82¢

Part 2 Answers will vary. Students' problems should fulfill the conditions set forth in the text. Sample responses are given.

- A variation on the alphabet name problem in Part **1a** with the values changed; for example, A = 26¢, . . . Z = 1¢.
- An addition problem such as Stretch Your Thinking exercise 5.

Assessment Rubric

3 The student accurately uses the given values for letters of the alphabet to find the value of various names; finds several names that have a value greater than $2.44; and writes a problem that meets all the criteria and finds the correct solution(s) for it.

2 The student uses the given values for letters of the alphabet to find the value of various names but may make a few mistakes in calculations; is able to find a name that has a value greater than $2.44; and writes a problem that meets some of the criteria and finds solution(s) for it.

1 The student is able to use the given values for letters of the alphabet to find the value of a few names but makes many mistakes in calculations; is not able to find a name that has a value greater than $2.44 without help; and cannot write a problem that meets the criteria and/or the solution(s) is/are not correct.

0 The student is unable to substitute given values for letters of the alphabet and shows little understanding of the concept; neither writes a problem that meets the criteria nor finds correct solution(s).

Name ____________________

6 MONEY PUZZLES

Wrap It Up

What's in a Name?

Part 1

Use these values for letters of the alphabet.

A	1¢	**J**	10¢	**S**	19¢
B	2¢	**K**	11¢	**T**	20¢
C	3¢	**L**	12¢	**U**	21¢
D	4¢	**M**	13¢	**V**	22¢
E	5¢	**N**	14¢	**W**	23¢
F	6¢	**O**	15¢	**X**	24¢
G	7¢	**P**	16¢	**Y**	25¢
H	8¢	**Q**	17¢	**Z**	26¢
I	9¢	**R**	18¢		

a. What is your name worth?

b. If your name was valued in money, which name would you want?

If you are a girl:	If you are a boy:
Millicent	Steven
Suzanne	Zachary
Rosalie	Mitchell

c. Use a telephone book to help you. Find a name that has the greatest value. Write down the page number on which you find the name. Try to top this: Walter Zytynsky. ($2.44)

Part 2

Make up your own problem in which letters are worth amounts of money. Solve your problem. Then share it with a partner.

2007

January

S	M	T	W	T	F	S
	1	2	3	4	5	6
7	8	9	10	11	12	13
14	15	16	17	18	19	20
21	22	23	24	25	26	27
28	29	30	31			

February

S	M	T	W	T	F	S
				1	2	3
4	5	6	7	8	9	10
11	12	13	14	15	16	17
18	19	20	21	22	23	24
25	26	27	28			

March

S	M	T	W	T	F	S
				1	2	3
4	5	6	7	8	9	10
11	12	13	14	15	16	17
18	19	20	21	22	23	24
25	26	27	28	29	30	31

April

S	M	T	W	T	F	S
1	2	3	4	5	6	7
8	9	10	11	12	13	14
15	16	17	18	19	20	21
22	23	24	25	26	27	28
29	30					

May

S	M	T	W	T	F	S
		1	2	3	4	5
6	7	8	9	10	11	12
13	14	15	16	17	18	19
20	21	22	23	24	25	26
27	28	29	30	31		

June

S	M	T	W	T	F	S
					1	2
3	4	5	6	7	8	9
10	11	12	13	14	15	16
17	18	19	20	21	22	23
24	25	26	27	28	29	30

July

S	M	T	W	T	F	S
1	2	3	4	5	6	7
8	9	10	11	12	13	14
15	16	17	18	19	20	21
22	23	24	25	26	27	28
29	30	31				

August

S	M	T	W	T	F	S
			1	2	3	4
5	6	7	8	9	10	11
12	13	14	15	16	17	18
19	20	21	22	23	24	25
26	27	28	29	30	31	

September

S	M	T	W	T	F	S
						1
2	3	4	5	6	7	8
9	10	11	12	13	14	15
16	17	18	19	20	21	22
23/30	24	25	26	27	28	29

October

S	M	T	W	T	F	S
	1	2	3	4	5	6
7	8	9	10	11	12	13
14	15	16	17	18	19	20
21	22	23	24	25	26	27
28	29	30	31			

November

S	M	T	W	T	F	S
				1	2	3
4	5	6	7	8	9	10
11	12	13	14	15	16	17
18	19	20	21	22	23	24
25	26	27	28	29	30	

December

S	M	T	W	T	F	S
						1
2	3	4	5	6	7	8
9	10	11	12	13	14	15
16	17	18	19	20	21	22
23/30	24/31	25	26	27	28	29